ECKART EHLERS & HERMANN KREUTZMANN
(EDS.)

HIGH MOUNTAIN PASTORALISM
IN NORTHERN PAKISTAN

# ERDKUNDLICHES WISSEN

SCHRIFTENREIHE FÜR FORSCHUNG UND PRAXIS
BEGRÜNDET VON EMIL MEYNEN
HERAUSGEGEBEN VON GERD KOHLHEPP,
ADOLF LEIDLMAIR UND FRED SCHOLZ

HEFT 132

FRANZ STEINER VERLAG STUTTGART
2000

ECKART EHLERS
& HERMANN KREUTZMANN (EDS.)

# HIGH MOUNTAIN PASTORALISM IN NORTHERN PAKISTAN

INCLUDING 20 PHOTOGRAPHS AND 36 ILLUSTRATIONS

FRANZ STEINER VERLAG STUTTGART
2000

Die Deutsche Bibliothek - CIP Einheitsaufnahme
**High mountain pastoralism in Northern Pakistan** / Eckart Ehlers &
Hermann Kreutzmann (eds.). - Stuttgart : Steiner, 2000
  (Erdkundliches Wissen ; H. 132)
  ISBN 3-515-07662-X

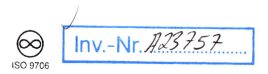

Jede Verwertung des Werkes außerhalb der Grenzen des Urheberrechtsgesetzes ist unzulässig und strafbar. Dies gilt insbesondere für Übersetzung, Nachdruck, Mikroverfilmung oder vergleichbare Verfahren sowie für die Speicherung in Datenverarbeitungsanlagen. Gedruckt auf alterungsbeständigem Papier. © 2000 by Franz Steiner Verlag Wiesbaden GmbH, Sitz Stuttgart. Druck: Druckerei Peter Proff, Eurasburg.
Printed in Germany

# PREFACE

The collection of articles presented in this volume is meant to be a focussed documentation of research on the problems and potentials of pastoralism in Pakistan's Karakorum Mountains. Centered around the general question of high mountain ecology and economy, the following case studies are results of empirical research that has been carried out within the Pakistan-German Research Project "Culture Area Karakorum" (CAK). This project, generously sponsored and supported by the Deutsche Forschungsgemeinschaft (DFG), has generated a wide range of new insights into the complicated interactions between nature and society in the harsh environment of northern Pakistan's high mountain belt. This compendium concentrates exclusively on aspects of traditional and modern forms of montane agriculture and animal husbandry and their combined potentials for a sustainable future development of those fragile ecosystems into which they have been and continue to be embedded.

Editors and authors of this volume dedicate this book to the memory of

CARL TROLL (1899–1975)

on the occasion of the 100th anniversary of his birthday on December 24th, 1999. In this sense, editors and authors would also like to see the following case studies in continuation of German geographical and comparative research on high mountains, a field to which Carl Troll has contributed so eminently.

Eckart Ehlers – Hermann Kreutzmann
Bonn/Erlangen, November 1999

# CONTENTS

High mountain ecology and economy: potential and constraints .................... 9
    ECKART EHLERS & HERMANN KREUTZMANN

Animal husbandry in domestic economies: organization, legal aspects and
    present changes of combined mountain agriculture in Yasin ................... 37
    GEORG STÖBER & HILTRUD HERBERS

Coming down from the mountain pastures: decline of high pasturing and
    changing patterns of pastoralism in Punial ............................................... 59
    REINHARD FISCHER

Pastoralism in the Bagrot: Spatial organization and economic diversity ........ 73
    ECKART EHLERS

Livestock economy in Hunza: societal transformation and pastoral practices  89
    HERMANN KREUTZMANN

Pastoral systems in Shigar/Baltistan: communal herding management and
    pasturage rights ...................................................................................... 121
    MATTHIAS E. SCHMIDT

Pastoral management strategies in transition: indicators from the Nanga
    Parbat region (NW-Himalaya) ................................................................ 151
    JÜRGEN CLEMENS & MARCUS NÜSSER

Why are mountain farmers vegetarians? Nutritional and non-nutritional
    dimensions of animal husbandry in High Asia ......................................... 189
    HILTRUD HERBERS

Addresses of Contributors ............................................................................. 211

# HIGH MOUNTAIN ECOLOGY AND ECONOMY POTENTIAL AND CONSTRAINTS

ECKART EHLERS & HERMANN KREUTZMANN

## 1. SUSTAINABLE AGRICULTURE IN A HARSH ENVIRONMENT

Development experts in general and protagonists of improved livelihood conditions in high mountain regions in particular tend increasingly to advocate the concept of sustainability. The creation and preservation of a stable socio-economic environment based on local resources remains to be a central focus of sustainable development. In an environment devoid of favourable conditions for productive crop production and animal husbandry two aspects are prominent in the discussion about future prospects:
- first, the strategic aim of an agricultural system which provides the basics for a growing population on a long-term basis without over-exploitation of available natural resources.
- second, the development of the specific potential of high mountain regions for the combination of production strategies in different ecological zones (Fig. 1).

Both approaches are rather restricted towards high mountain regions in developing countries (cf. JODHA, BANSKOTA & PARTAP 1992, JODHA 1997), whereas the respective areas in industrialized countries are characterized by economies preferably dependent on the secondary and tertiary sectors (cf. LICHTENBERGER 1979, UHLIG 1995). It is still an open question which future role and function high mountain agriculture may play in the Third World when it is embedded in an overall market economy where low sustainability levels are attributed to its competitiveness (RIEDER & WYDER 1997).

For the purpose of this study it is more important to note that obviously no role models for a sustainable high mountain agriculture or other harsh environments are provided from those countries where the major donors in development aid and assistance are located. However, replicable examples have been derived from case studies in remote regions of Africa, Asia and Latin America (cf. for example CASIMIR 1991, KRINGS 1991, RICHARDS 1985) and rarely touch high mountain environments. Besides, the great majority of studies which were investigating experiences with sustainable agriculture have been focusing on the so-called traditional sector in a rather isolated manner. They neglected very often the fact that agriculture is only part of a much broader rural economy and part of an often intricate system of labour division.

In view of the fact that high mountain agriculture is in most cases more than agriculture per se, it may be appropriate to stress from the very beginning that

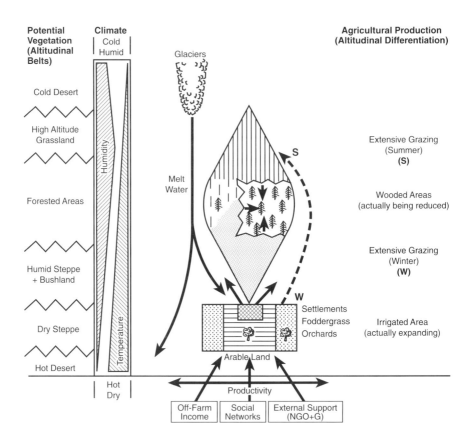

Fig. 1: Vertical arrangement of natural vegetation and agricultural productivity
(Source: M. Winiger 1996)

mountain livelihoods are mostly based on a combination of various income sources: agriculture, pastoralism, forestry, handicrafts, various services and trades and – more recently – local industries, tourism or mountaineering provide a diversity of household incomes in various combinations. Such a statement does not imply that the majority of mountain dwellers may still be subsistence farmers or are depending in their majority on agricultural practices. But for a substantial number of mountain dwellers agriculture remains to be an enterprise based on locally available resources and an economy providing a significant share to the household income. The complex set of contributions to the household income cannot disguise the fact that the agricultural sector functions as a safety valve when other occupations fade and non-agrarian enterprises fail. Thus it seems worthwhile to devote a separate study to high mountain agriculture and certain aspects of it.

## 2. HIGH MOUNTAIN AGRICULTURE

In contrast to profitable agricultural practices in highly productive core regions, often symbolized by mono-cultural cultivation patterns and/or farming practices involving substantial technological and financial inputs, adaptive strategies in the peripheries of the ecumene are regularly characterized by the combination of different resources. The combination of crop-farming and livestock-keeping seems to be the basis of almost any high mountain agriculture. Both sectors interrelate and are linked (Fig. 2). The livestock provides substantial amounts of animal manure to fertilize the fields, while a certain part of the cultivated lands has to be reserved for the fodder production. Parcels of land solely reserved for animal husbandry might carry alfalfa crops (*Medicago sativa*) or comprise irrigated or non-irrigated meadows. Grass and lucerne cultivars are cut several times per growing season. In addition, other cultivated plants contribute consumable parts such as leaves, residue or straw for the livestock. Fodder is stored to support the livestock during seasons when natural grazing is not available or too limited. Thus in a high mountain agriculture both sectors are interdependent which led RHOADES & THOMPSON (1975) to suggest the terminus of "mixed mountain agriculture". In their search for an appropriate and universal terminology they refrained from projecting concepts derived from European experiences onto non-

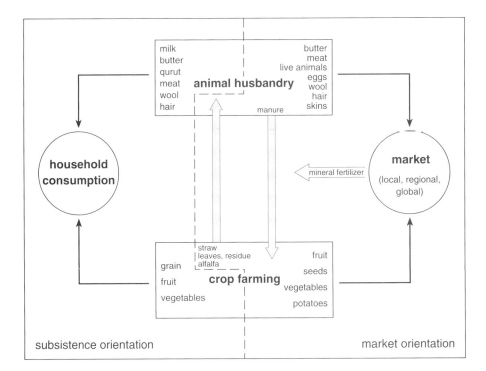

Fig. 2: Exchange relationships in high mountain agriculture

European situations and locations. The "mixed" mountain agriculture addresses a set of different activities and sources of income which are interdependent in resource utilization strategies, the management of farms and the division of labour. This somewhat neutral ascription of a dominant human intervention in high mountain regions functions as an umbrella for specialized and regionally differentiated articulations. Presently we find its application in High Asia when the International Centre for Integrated Mountain Development (ICIMOD, based in Kathmandu) addresses agricultural practices in the Hindukush-Karakoram-Himalaya as "mixed crop farming systems" (ICIMOD 1998).

The combination of crop cultivation and livestock-keeping is more widespread and complex in the Old World and especially in the high mountain regions of Asia than anywhere else (cf. BISHOP 1990, CROOK & OSMASTON 1996, EHLERS 1976, 1980, BALLAND 1988, JENTSCH & LIEDTKE 1980, PRICE & THOMPSON 1997, STEVENS 1993, RATHJENS, TROLL & UHLIG 1973) as shall be shown below.

The concept of verticality, sometimes addressed as vertical control, was developed in the Central Andes. Mainly it was projected onto tropical mountains of other continents and it highlights the layered structure of agricultural potential along the gradient of the slope creating specialized production and exchange of goods between ecological zones (BRUSH 1976a, b, BRUSH & GUILLETT 1985, FORMAN 1978, MAHNKE 1982, MURRA 1975). "Classifications of types of verticality have been developed to make the extension of the concept more useful. BRUSH (1976b) distinguishes "archipelago" types (the different zones are controlled by the same group are and non-contiguous and often separated by great distances), the "compressed" type (the zones controlled by the same group are contiguous), and the "extended" type (the zones are contiguous but controlled by different groups, who engage in barter or market sale); FORMAN (1978) adds a "mixed" type, which includes access by one group to multiple zones both by direct control and through market exchange." (ORLOVE & GUILLET 1985, 9). Different crops are cultivated in a number of layers. Traditionally livestock-keeping was mainly restricted to the upper zone (*tierra helada*), lacking the seasonal migrations of the above-mentioned high mountain agriculture. Its interrelationship with the crop cultivation complex is weaker as the manure producers (= livestock) are normally not moving to the village lands for fertilization. Basically periodical herd migrations and the seasonal availability of natural grazing distinguishes tropical and subtropical strategies. In recent years the cultural-geographical and anthropological discussion has stressed that the compressed type of vertical control resembles very much mixed mountain agriculture or what is grasped as "Alpwirtschaft" (BRUSH & GUILLET 1985, GUILLET 1983, ORLOVE & GUILLET 1985, RHOADES & THOMPSON 1975).

The historical context, different influences and research perspectives directed towards mountainous resources were made responsible for the classificatory stalemate. Consequently, attempts to transfer the verticality approach to Inner Asian mountain systems (CASIMIR & RAO 1985) were introduced as a means to compare the utilization strategies in different ecosystems and connected production systems. The emphasis in the following discussion is put on animal husbandry as an integral part of a more complex socio-economic system. Livestock-keeping and/or crop cultivation are two aspects of primary production while a varying number of non-agrarian activities contribute to household-based enterprises and incomes in addition. The share and degree of primary production may vary significantly in space and time.

## 3. PASTORAL PRACTICES IN HIGH MOUNTAIN REGIONS

The spectrum of animal husbandry in high mountain regions transgresses its sole incorporation into the complex of sedentary agriculture. Pastoral practices solely devoted to livestock-breeding are found in high mountain regions as well. For the purpose of clarification it seems advisable to devote some attention to the different practices which can be identified in high mountain regions. Guiding principle in this approach is neither the search for essential types nor the regionalization of utilization strategies. Categories are solely introduced for the purpose of stream-lining practices which incorporate different adaptive and socio-economic livelihood strategies.

In the general discussion of pastoralism classificatory concepts derived from European experiences are projected onto other areas. Principally there is nothing detrimental about applying terms such as "Almwirtschaft/Alpwirtschaft" from the European Alps or "Transhumance" from the riparian mountain regions of the Mediterranean Sea to phenomena which follow a similar practice when the contents is specified. But very often the application of these terms is neclecting local socio-cultural conditions. Secondly, sometimes models of such practices are presented as "typical" which, however, describe cases of only recent existence. A diagram from ARBOS (1923, 572) depicting the movement of mountain farmers and their livestock in the French Alps (Tarentaise) is typical. It was reproduced as the only model in the textbook of PRICE (1981, 413) more than half a century after its conception. A second somewhat younger case is the diagram created by PEATTIE in 1936 in which he presented a time-space model of the seasonal migration by people practicing "Almwirtschaft" in the Val d'Anniviers of the Swiss Alps. Since the 1950s this practice has ceased, but the diagram was published again and again (GRÖTZBACH 1982, 10, 1987, 65, 1988, 27). The latest appearance occurred in GRÖTZBACH & STADEL (1997, 26). The reproduction again and again support the impression of stagnant features in high mountain regions and less susceptibility to social change. On the other hand, we find a number of cases – evidence from High Asia will be presented in the contributions of this volume – where a wide spectrum of pastoral practices is very much alive and prominent in every-day life.

BEUERMANN (1967, 17–31) and SCHOLZ (1982, 2–8; 1992, 7–16) provided a consistent descriptive scheme of categories for the interpretation of non-stationary forms of animal husbandry. For the scope of this study three classes/categories are introduced which are linked to the utilization of high mountain pastures by distinguishing mobility patterns, socio-economic organization and property rights (Fig. 3):

(i) *Mountain Nomadism*: The nomadic economy and labour activities are predominantly based on animal husbandry. Mixed herds are composed of sheep and goats, cattle/yaks for livestock production and camels, horses and donkeys mainly for transport of tents, household goods and utensils. The whole group covers great distances between lowlands and highlands during their seasonal migrations towards suitable and accessible pastures. The mobile communities show strong kinship relations. As a rule they distinguish themselves from their neighbours and business partners as a social group of livestock proprietors and traders. Nomads utilize pastures for which they claim the rights of access based on customary law, nevertheless grazing taxes are levied and paid to the state or private persons. Such are the business relations for pastures in addition to barter trade with farmers for basic goods such as grain. Traditionally mountain nomads have engaged only in very few side activities beyond animal husbandry such as transportation, trade, services and other businesses while crop cultivation was no practice attributed to nomads. The absence of permanent settlements and village lands resulted in a mobile society in which movable property including tents and yurts was characteristic and provided shelter in the grazing grounds. Both traditions have been changed quite tremendously in all societies in recent years (Photo 1). Planned and forced sedentarisation of nomads, the introduction of permanent

Photo 1: Bakarwal nomads shifting camp from the Nanga Parbat region down the Indus Valley. Some members of the grazing community move ahead of the flocks to set up camp. The Karakoram Highway provides a comfortable route for part of the migration (H. Kreutzmann, September 27, 1995).

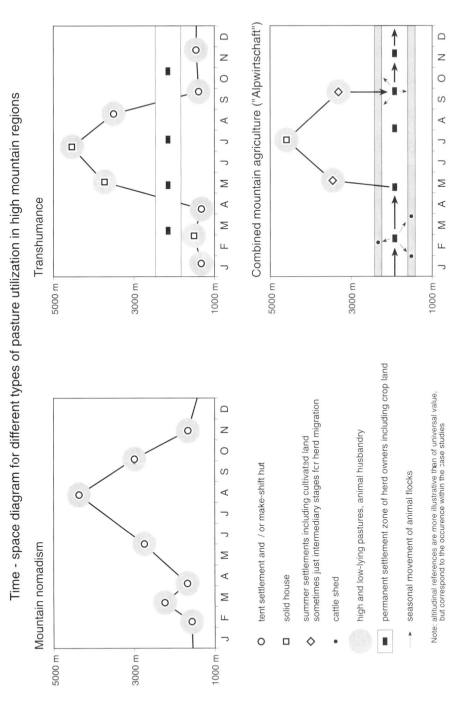

Fig. 3: Time-space diagram for different types of pasture utilization in high mountain regions

winter camps, agrarian reforms and socio-economic change have resulted in adjusted and comparatively confined migration cycles. All factors have contributed to a controlled mobility with features of permanency such as houses and stables in a community settlement. The expansion of crop cultivation and village lands, the reduction of available space and the progression of bureaucracy have limited the rangelands and pastures accessible for nomads. Territorial, political and private delineation of boundaries increased the phenomenon of "closed frontier nomadism" (SHAHRANI 1979). For anthropological and cultural-geographical contributions in the spectrum of mountain nomadism cf. BALLAND 1988, BARFIELD 1993, BARTH 1956a, b, BHASIN 1996; CASIMIR 1991, EHLERS 1976, 1980, GOLDSTEIN & BEALL 1990, GOLDSTEIN, BEALL & CINCOTTA 1990, HÜTTEROTH 1959, 1973, JEST, 1973, JETTMAR 1960, KHATANA 1985, KREUTZMANN 1995a, 1998b, LANGENDIJK 1991, RAO 1992, SCHLEE 1989, SCHOLZ 1982, 1991, 1992, 1995, 1999, SHASHI 1979, E. STALEY 1966, STÖBER 1978, 1979, 1988, TUCKER 1986, UHLIG 1962, 1965, 1973.

(ii) *Transhumance*: Described about a century ago as a regional pastoral practice in Southern France, the term transhumance has gained many connotations and global application in recent years. Sometimes it is used in a wide sense synonymous for pastoralism and nomadism as a comprehensive concept in Anglo-American publications, sometimes it describes pastoral practices linked to certain ethnic groups while the narrow interpretation with a focus on flocks prevails in non-English language usage (BLACHE 1934, RINSCHEDE 1979, 1988, 97–98). The latter addresses some features of the livestock-related agrarian activities originally observed in the riparian states of the Mediterranean Sea. Transhumance (Fig. 3) involves seasonal migrations of herds (sheep and goats, cattle) between summer pastures in the mountains and winter pastures in the lowlands. In contrast to mountain nomadism the shepherds of a migrating team are not necessarily that strongly affiliated with each other to form a group of relatives managing their own resources. The shepherds serve as wage labourers hired by the livestock proprietors on a permanent basis. As a rule the shepherds are neither related to them, nor do they have livestock of their own. The proprietors of the flocks can be farmers or non-agrarian entrepreneurs. Managementwise the year-round migration between suitable grazing grounds is independent from other economic activities of the proprietors. In case they are farmers, their farm management and agricultural activities are not interrelated with their livestock. Nevertheless sometimes proprietor farmers provide shelter and grazing on their fields after harvest or on meadows. Usually common property pastures are utilized in the mountains while customary rights or contracts with residents in the lowlands establish the winter grazing conditions. Shepherds live in mobile shelters (tents, carts etc.) or in permanent houses provided to them. Transhumance of this kind is found in mountainous regions of all continents (cf. RINSCHEDE 1988, 99–100), there is no general trend of decline to be observed although its share in pastoralism varies in wide ranges. For contributions to the scholarly discussion of transhumance cf. BEUERMANN 1967, CHAKRAVARTY-KAUL 1998, GRÖTZBACH 1972, 118–120, HOFMEISTER 1961, JETTMAR 1960, KHATANA 1985, MANZAR ZARIN & SCHMIDT 1984, RINSCHEDE 1979, 1988, SNOY 1993, UHLIG 1995.

(iii) *Almwirtschaft/Alpwirtschaft or Combined Mountain Agriculture*: The terms "Alm" (Austrian, Bavarian), "Alp" (Swiss) and "alpage" or "élevage avec estivage" (French) refer to the high pastures as a characteristic and epigrammatic feature of the European Alps. In this narrow sense only one seasonal aspect of high mountain agriculture is addressed, the utilization of alpine grazing grounds by mountain farmers during summer. At the same time "Alp/Alpen" is a term describing the whole mountain system or a general mountain range. Consequently "Alpwirtschaft" could be understood as the specific form of agriculture prevalent in the European Alps. In the wider sense it adresses what RHOADES & THOMPSON (1975) understand as mixed mountain agriculture and what GUILLET (1983, 562-563) introduced as the adaptive strategy of "Alpwirtschaft". While they identify it as "agropastoral transhumance", this viewpoint is questionable. There are a number of differences between both strategies in livestock-keeping. In Alpwirtschaft (Fig. 3) the proprietors of the flocks are residents of the home-

Photo 2: The summer settlement of Bari (3300 m) in the Ishkoman Valley (Ghizer District). The huts and animal sheds are constructed of locally available stones while the cone-shaped roof is assembled of juniper branches. In these settlements the lactating herd animals are kept during the night where they are milked in the evening and in the morning before being brought daily to sometimes quite far away pastures. Shepherds process the milk and care for the new-born animals during the day here. In addition they collect firewood and construction timber as well as animal manure for transport to the homesteads (H. Kreutzmann, September 3, 1990).

Photo 3: The pasture settlement of Boiber (3400 m) above Abgerch in Gojal (Hunza). The flocks of sheep and goats are kept overnight in a pen and are brought to the daily pasture in higher locations by female shepherds while ailing and weak animals remain in Boiber where the other women spend the day with the processing of milk, preparation of food, collection of firewood and manure etc. (H. Kreutzmann July 8,

steads in the valley grounds. They initiate and control the organization and management of the grazing cycles as a measure to increase agricultural productivity and livestock numbers in a given territory. The example of the Hindukush-Karakoram-Himalaya region shows that not only the herd sizes can be increased by incorporating high pastures into the domestic economy. But at the same time the quality of natural grazing in the high pastures has been estimated as double to quadruple of that in the lower zones of the arid mountain valleys (SHEIKH & KHAN 1982, quoted in STREEFLAND, KHAN & VAN LIESHOUT 1995, 85). On the one hand, the agricultural production in the homestead is strongly linked to the livestock sector by growing grass and storing hay for the winterly provision of fodder. During the winter period the flocks are kept in stables or in the open close to the permanent dwellings of the mountain farmers. The shepherds are traditionally members of the extended family although in recent years a tendency for the employment of hired professional labourers can be observed. With growing job opportunities it becomes more difficult for mountain farming households to provide the manpower especially during the summer season when the agricultural workload is high and other monetary job opportunities might be available. Thus it is a quite common feature that households pool their livestock and send the herds with a trusted person or a hired professional to the summer pastures which are mainly part of common property (Photos 2 & 3). For classical and recent contributions to the discussion of Alpwirtschaft cf. AZHAR-HEWITT 1999, BHASIN 1996, BISHOP 1990, BUTZ 1996, FRÖDIN 1941, HERBERS 1998, HERBERS & STÖBER 1995, HEWITT 1989, ITURRIZAGA 1997, JEST 1974, 1975, KARMYSCHEVA 1969, 1981, KREUTZMANN 1986, 1999, KUSSMAUL 1965, MUKHIDDINOV 1975, NÜSSER 1998, 1999, NÜSSER & CLEMENS 1996, PARKES 1987, SCHWEIZER 1985, SNOY 1993, STADELBAUER 1984, STALEY 1966, 1969, 1982, STEVENS 1993, TROLL 1973, UHLIG 1976, 1995.

As will be seen in the contributions to this collection of articles, the direction of change and the related adjustments to households' division of labour and responses for agricultural activities are a complex set embedded in the overall economy. Pastoral activities can only be understood as part of the domestic economy. Some authors attributed to the livestock sector such an importance that they identified the utilization of pastures as the key indicator for a cultural-geographical typology of high mountain systems (GRÖTZBACH 1980). This view forms a cultural-geographical response to Carl TROLL's (1962, 1972, 1975) three-dimensional landscape classification and comparative high mountain geography which was dominated by a geo-ecological structuring. The essential features addressed were ecological zones. From the point of human intervention the time-space relationship between crop farming in suitable locations and animal husbandry in the highest available locations above the tree-line was perceived as the most important aspect for typological purposes. GRÖTZBACH (1980, 273) differentiated high mountain regions world-wide by assessing the degree of intensity in mountain livestock-keeping and pasture utilization. In his overview, the extra-tropical mountain ranges of the Old World are identified as regions of medium intensity (European Alps, Pyrenees, Abruzzis, Carpathians, and the high mountain ranges of the former Soviet Union) and high intensity (Atlas, Himalayan system, and the high mountain ranges of the Middle East). The contributions to this compendium are located in the latter group. Nevertheless SCHWEIZER (1984, 36) has pointed out that these classificatory specifications are not leading to an identification of relevant research topics. They belong rather to the ivory towers of academic institutions. It has been a long way to discuss sustainable agriculture in the context of high mountain research. The changing perspectives upon high mountain pastoralism might be illustrated, with a few sketches.

## 4. STATE OF THE ART: DEVELOPMENT OF RESEARCH ACTIVITIES IN THE FIELD OF HIGH MOUNTAIN PASTORALISM

Initial impulses for high mountain research stem from studies in the European Alps which have functioned as a model region for empirical studies and for conceptual approaches at large. This also holds true for one of the earliest attempts to analyze the closer interactions between society and montane environments: Jules BLACHE's "L'Homme et la Montagne" (1934). Although covering a worldwide perspective, this book – strangely enough – not only omits the Himalaya-Hindukush region, but focusses remarkably strong on alpine and mediterranean Europe. It may be therefore allowed to say that it was especially Carl TROLL who has broadened the perspective by interpreting the existing literature in a comparative view and by supporting research activities outside Europe. In the early stages high mountain research remained to be heavily concentrated on ecological features. With a considerable time gap cultural geographers joined in and developed their own classifications. Research topics have been e.g. demographic pressure (DE PLANHOL 1968, SKELDON 1985), the utilization of high altitude pastures (EHLERS 1976, 1980, GRÖTZBACH 1980, RATHJENS 1973, UHLIG 1976, TROLL 1973), high mountains as human habitat (BLACHE 1934, BRUSH 1976, GUILLET 1983, GRÖTZBACH 1982, PANT 1935, PEATTIE 1936, RHOADES & THOMPSON 1975, SCHWEIZER 1981, UHLIG 1984). A comparative cultural geography of high mountain regions (GRÖTZBACH 1976, JENTSCH 1977, RATHJENS 1968, 1981, 1982, RATHJENS, TROLL & UHLIG 1973, SCHWEIZER 1984, VEYRET & VEYRET 1962) as well as aspects of human impacts on mountains (ALLAN, KNAPP & STADEL 1988, PRICE 1981, SOFFER 1982), moved into the centre of attention. The utilization of high pastures and changes over time played a vital role in those studies. In the early eighties the complex system of ecology and economy, as well as the antagonism of self-reliance and dependency in mountain regions, was highlighted and led to a number of studies. Pastoralism formed one of the three major fields of interdisciplinary interest along with a focus on forest depletion and water utilization. National research efforts, such as the report on "the transformation of Swiss mountain regions" (BRUGGER ET AL. 1984) and the supra-national UNESCO programme "Man and Biosphere", furthered the knowledge of environmental, economic and cultural processes in the Alpine belt and initiated intensified scientific activities in this field. An overview of the results of comparative high mountain research was presented by HEWITT (1988), RATHJENS (1988), UHLIG (1984) and UHLIG & HAFFNER (1984), the latter date the beginning of research interest in "Alpwirtschaft" back to the first decade of the 20$^{th}$ century (SIEGER 1907).

Voices could be heard demanding more integrated research efforts (CHAUBE 1985, LAUER 1984, 1987, SINGH & KAUR 1985), as well as expecting an increased attention towards changing traffic infrastructures, accessibility (ALLAN 1986) and the integration of mountain regions in the global economy (KREUTZMANN 1993a, 1995b). The attempt to broaden the view for a holistic approach led to greater emphasis on defining common research topics. Again the utilization of high pastures featured prominently as the transformation in the household economy affected all sectors of mountain agriculture. The changing division of labour, outmigration, tourism and non-agrarian employment affected the availability of pastoral workforce significantly; at the same time the impact of education was strongly felt. The complexity and interdependency of mountain economies was acknowledged and left the simplistic distinction in tradition and modernity (BISH-

OP 1990, KREUTZMANN 1989). The perceived need for multidisciplinary perspectives was increased by the attempt to view mountain regions as part of a highland-lowland exchange system incorporating the dimension of a world economy in which mountain communities participate on different levels. Such a view was acknowledged and easily accepted in respect of mountaineering, trekking and recreation tourism, but less obvious in studies focussing on culture and nature conservation. The protection of endangered ecological niches and ecotopes, as well as the provisional function for natural resources, has been highlighted in recent years.

In an exemplary manner research deficits were exposed by questioning the validity of the so-called "Theory of Himalayan Environmental Degradation" (IVES 1987) and realizing the "uncertainty on a Himalayan scale" (THOMPSON, WARBURTON & HATLEY 1986). Methodological shortcomings, particularistic approaches and isolated mono-disciplinary studies resulted in the equivocal statement of the "Himalayan Dilemma" (IVES & MESSERLI 1989), which revealed many significant research deficits and a multitude of development problems in a particular mountain belt. Nevertheless, shortcomings in the human sphere were more than obvious.

Presently mountain research is directed towards a continuation along the path of homogenization of methods and problem-targetted studies with a regional focus (MITCHELL & GUILLET 1993, KREUTZMANN 1998a, STELLRECHT & WINIGER 1997, STEVENS 1993). Nevertheless, the integration of local scholars, activists and professionals remains to be a desideratum not only in defining research topics but in cooperative research programmes as well. As a contribution towards the "Rio+5 Conference" in New York in 1997, a further combined effort resulted in the publication of "Mountains of the World. A global priority" (MESSERLI & IVES 1997) in which efforts for sustainable development embodied in local communities are advocated. It is before the specific background of this publication and the wealth of applicable suggestions that it contains the following plea: to conserve and preserve the upper reaches of the alpine and subalpine mountain environments as an indispensable token of future sustainability and as a viable protective resource for generations to come. Such a plea should not be misunderstood as a "benevolent recommendation" from far away and from a representative of a society which can afford protection of natural environments almost like a luxury item. The contrary is the case: high mountain environments in industrial countries like the European Alps or the North American Rocky Mountains have suffered over centuries from ecological mismanagement and brutal economic exploitation – partly as a result of the historical poverty of the mountain dwellers, partly because of the greed of the forelanders in regard to the montane forests, water energy, and mineral resources. The results of such short-sighted exploitation have, however, been disastrous in the long run. Therefore: it is experience that lies behind these recommendations and that tries to ensure future sustainability.

Especially the volume edited by MESSERLI & IVES (1997) contains a wealth of suggestions and strategies in regard to the sustainable development and the role

of pastoralism in montane ecosystems. Thus JODHA (1997, 313–314) argues that "diversified, interlinked, land-based activities; folk agronomy involving measures with low land intensity and low (local and affordable) input regimes..." have been maintained by "traditional farming systems", including the applications of indigenous forms of agroforestry or periodic uses of certain vegetation belts. The protection of those ecosystems is not only essential for the preservation of watersheds and soils, but also for the reduction of natural hazards such as flooding, avalanches, land slides, or debris flows. It is therefore more than just an unwarranted demand and recommendation by scientists from "rich" countries when they are demanding as key themes for action to sustain the future of mountain communities and their environments:
- mountain-lowland interactions;
- sustainable agriculture and forest management and development;
- economic and technological changes and mountain communities;
- appropriate institutional arrangements.

Such and other recommendations (European Conference on Environmental and Societal Change in Mountain Regions: Global Change in the Mountains. Conclusions and Recommendations. Oxford, March 1998; edited by PRICE 1999) are almost necessarily of special importance to those fragile mountain ecosystems that are occupied by pastoralism and mountain forestry. In this context it should be clear that the demands of "sustainable agriculture" and "forest-management" alone are not enough. Experience shows that local farmers are well aware of ecological necessities. It is therefore essential that local, regional and national governments and their representatives understand and play their key roles in the implementation of these and other strategies.

In spite of the fact that the attention drawn to pastoral practices in the volumes or articles edited by MITCHELL & GUILLET (1994), STELLRECHT (1998), STELLRECHT & WINIGER (1997) or published by STEVENS (1993) and others are rather limited, it nevertheless holds true what Carl RATHJENS (1988, 140) stated: "The economy of high pastures belongs to the larger field of grazing and of seasonal migrations of man and cattle to high mountains. Therefore these facts have met with a growing interest in many mountains of the world, even though their importance for tropical high mountains is not yet fully understood."

Since this observation about a decade ago the study of pastoral practices has been incorporated into the research design of the "Culture Area Karakoram" (CAK) project of which results are presented in this compendium. In High Asia we generally observe a still important share of animal husbandry in the agricultural sector which remains to function as a basic needs provider although its share in contributing to the livelihood of mountain dwellers is decreasing. Connected with this is the perception that significant transformations occur in the field of land use strategies and the utilization of natural grazing. These changes can only be understood when seen in the wider context of high mountain economies. Therefore, research activities are rarely focused on the livestock sector alone. This statement holds true for the regional studies presented here.

## 5. AIMS AND FOCUS OF THIS COMPENDIUM

The investigations are challenging a developmentalist paradigm which governed the first four decades of development cooperation based on insufficient knowledge and expertise about local and regional conditions. Having made this state-

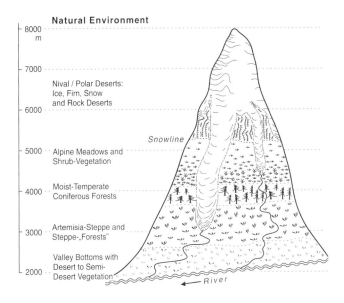

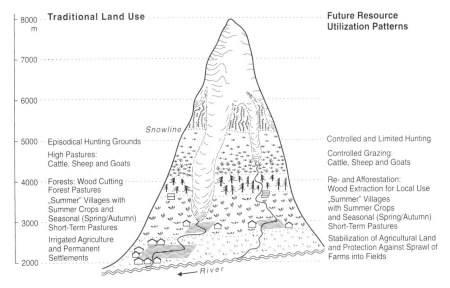

Fig. 4: Correspondence between physical environment, present land use and future resource utilization patterns

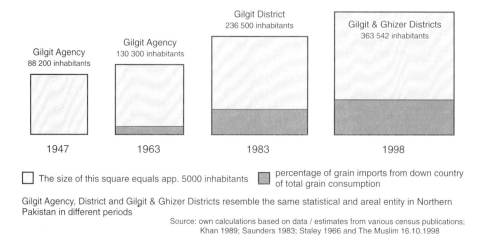

Fig. 5: Northern Pakistan: population growth and the development of food dependency

ment, the dualism of tradition and modernity is questioned as well as the modernization paradigm itself and a uni-directional path of development. Consequently popular linkages such as the interrelationship of population growth with land degradation and/or the unrestricted exploitation of natural resources have to be analyzed and discussed in the framework of socio-economic change. The impact of a changing accessibility through the provision of roads and motorable tracks is a key issue for the incorporation of peripheral mountain societies into the lowland economies and the world market. Dependency on local natural resources might be reduced or increased. Out-migration affects the consumer patterns as well as the population structure of the residents within the mountains. Exchange relations have undergone and are undergoing tremendous transformations and affecting the societal setup. With the advent of improved transport and communication systems – in the case of Pakistan the Karakoram Highway (KKH) stays as the symbol of such a transition – external agents of change appear on the scene in a new dimension. Public and privately sponsored economic planning and innovative measures have resulted in a wide spectrum of development activities under the banners of basic democracies, integrated rural development, community basic services programmes, local bodies & rural development, rural support programmes etc. In recent years their contributions and interferences have significantly altered the traditional exchange patterns between farming households and respective markets for their products and consumer demands. Socio-political developments such as transformations in the administrative setup, the extension of traffic and social infrastructure for the provision of basic amenities affect everyday life, the organisation of households and the divison of labour (HERBERS 1998, HEWITT 1989, KREUTZMANN 1995b, MILLS 1996). The close interactions between harsh physical environments, increasingly unsustainable land uses and desirable potential forms of resource utilization are presented in Fig. 4 where ecological frame conditions are interpreted as a limiting, but not necessarily uni-laterally determining factor for human action.

Pastoral practices as introduced above have been affected and transformed in a similar manner as other fields of activity. Northern Pakistan has experienced this accelerated change since two decades (KREUTZMANN 1991, 1995c, STREEFLAND, KHAN & VAN LIESHOUT 1995). According to data from the population census offices the population has quadrupled since independence. Population growth continues and the dependency on external goods grows faster (Fig. 5). Nevertheless social and regional inequalities are tantamount to a complex pattern of socioeconomic and market participation which cannot be explained by unifactorial correlations. Taking pastoral practices as the focus of attention the case studies shed light onto the overall dynamics of change in the high mountain regions of Northern Pakistan. According to estimates by the administrative authorities pastures cover nearly half of the available land (Fig. 6).

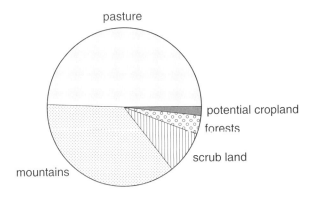

Source: Streefland, Khan & van Lieshout 1995, 83

Fig. 6: Land use differentiation in Northern Pakistan (Chitral and Northern Areas)

Institutions involved in rural development assess the requirement and availability of high pastures from previous estimates and observe a shift since the 1950s. The period immediately after independence is termed as a period when "pastures were still abundant" (STREEFLAND, KHAN & VAN LIESHOUT 1995, 85). The total number of livestock in the Northern Areas ranged around 1.12 million units of animal equivalents (equalling tropical livestock units) while natural grazing covered 3.6 million hectares (SHEIKH & KHAN 1982, 40–41). Cattle herds increased by a quarter between 1970 and 1990, during the same period a decrease of sheep and goats' numbers of 12% was stated. The Pakistan Livestock Census of 1986 states for the Northern Areas an animal herd composition of: 408,056 cattle, 979 buffaloes, 642,645 sheep, 938,897 goats, 4,743 horses, 25,371 donkeys, 703 mules and 358,998 poultry birds (AKRSP 1991: 5). Although all figures are rather vague and based on extrapolations, livestock experts support the statement that in the meantime overgrazing or underfeeding has reached a quota of app. 30%. Acknowledging this, STREEFLAND, KHAN & VAN LIESHOUT (1995, 85) claim that "pastures do not seem to be irrecoverably damaged". They attribute the shifts to a "conversion of lower pastures and temperate range pastures into cropped land, which itself is used for house construction. ... The trend to convert low yielding pastures into high yielding cropped land is therefore a favourable one ..." (ibid., 85–86). "Average animal feed self-sufficiency has come down to a meagre 6–9 months. An increase in fodder trade is the logical consequence." (ibid., 91). The conversion of natural pasture into irrigated farmland is seen as a strategy to counterbalance this trend and to increase overall productivity besides providing alfalfa, clover, straw and plant residue to the livestock.

## 6. CASE STUDIES

The contributions presented in this volume attempt to investigate these interrelated patterns and changes in the production system. Without exception they are based on extensive field-work to provide insight into the complexities of resource utilization in the Eastern Hindukush, Karakoram and the Western Himalaya.

All practices of animal husbandry as introduced above – mountain nomadism, transhumance, and combined mountain agriculture – are to be found in the study area (Fig. 7). The latter is dominating the present period. Mountain farmers combine the natural resources made available for summer grazing in the high pastures with their all-season agriculture in the arid valley bottoms. Seasonal migrations of herds and shepherds are required to connect the different production zones. Substantial shares of the irrigated lands are reserved for fodder production (meadows, field parcels for alfafalfa and clover) and residue from crop cultivation supports the feed requirements of the flocks in addition. Mainly cattle are kept throughout the year in the homestead while sheep and goats as well as yaks are removed during the cultivation period. This form of combined mountain agriculture is to be found all over the study area in the Eastern Hindukush, Karakoram, and Western Himalaya and has replaced mountain nomadism in certain regions such as the Hunza valley (KREUTZMANN 1989, 1996, SNOY 1993). Mountain nomadism has become a historical feature there while we still find nomadic Bakerwal, Pashtuns and Gujur in the Nanga Parbat region, the Kaghan Valley and the Eastern Hindukush respectively (GRÖTZBACH 1989, 6, LANGENDIJK 1991, SCHICKHOFF 1993, 179–184). The Gujur people cannot be termed mountain nomads alone as some of them have settled down in the mountain valleys and follow a similar form of combined mountain agriculture as

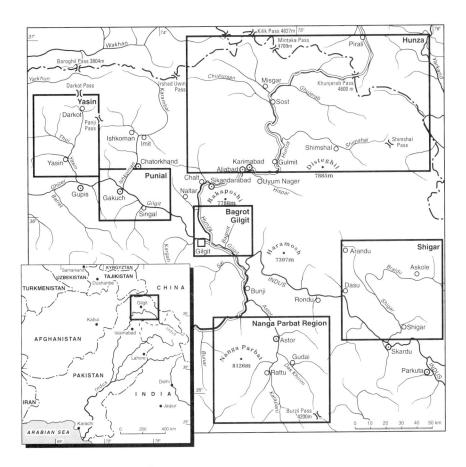

Fig. 7: Location of case studies

their neighbours. First Gujur migrations to the Ishkoman valley are recorded in 1910, other areas must have been visited much earlier. Some Gujurs have become shepherds for the local elite and afterwards their migrations resemble what has been described as transhumance above (BARTH 1956a, 1079–1089, 1956b, 76–78, JETTMAR 1960, 125, 129, KREUTZMANN 1994, 348, 1996, 257–262, LANGENDIJK 1991, SNOY 1965, 119–121, E. STALEY 1966, 334-344, J. STALEY 1966, 33–34, 96). Gujurs are mainly to be found nowadays in the western and southern part of the study area (Fig. 7). This brief discussion underlines the fact that regionalization of practices and their attribution to certain groups are obviously oversimplifying socio-economic activities which have been utilizing space differently and have been changing over time.

## 7. ACKNOWLEDGEMENTS

All case studies presented here have been executed within the framework of the *Deutsche Forschungsgemeinschaft* (DFG, German Research Council) sponsored research programme "Culture Area Karakorum" between 1989 and 1995. The authors and editors gratefully acknowledge the generous support extended to the participants of the project for fieldwork and archival studies. Without the support, cooperation and hospitality of public and private institutions, interview partners, experts and friends in Northern Pakistan, none of the contributions to this volume would have been feasible.

## 8. BIBLIOGRAPHY

Aga Khan Rural Support Programme (1991). A report on livestock in the Northern Areas. Gilgit.

ALIROL, P. (1979): Transhuming animal husbandry systems in the Kalinchowk region (Central Nepal): a comprehensive study of animal husbandry on the southern slopes of the Himalayas. Bern.

ALLAN, N. J. R. (1986): Accessibility and altitudinal zonation models of mountains. In: Mountain Research and Development 6, 185–194.

ALLAN, N. J. R., KNAPP, G. W. & C. STADEL (eds.) (1988): Human impact on mountains. Totowa, New Jersey.

ARBOS, P. (1923): The geography of pastoral life. In: Geographical Review 13, 559–575.

AZHAR-HEWITT, F. (1999): Women of the high pastures and the global economy: Reflections on the impacts of modernization in the Hushe valley of the Karakorum, Northern Pakistan. In: Mountain Research and Development 19 (2), 141–151.

BALLAND, D. (1988): Nomadic Pastoralists and sedentary hosts in the Central and Western Hindukush Mountains, Afghanistan. In: ALLAN, N. J. R., KNAPP, G. W. & C. STADEL (eds.): Human Impact on Mountains. Totowa, New Jersey, 265–276.

BARFIELD, T. J. (1993): The nomadic alternative. Englewood Cliffs.

BARTH, F. (1956a): Ecological Relationships of Ethnic Groups in Swat, North Pakistan. In: American Anthropologist 58 (6), 1079–1089.

BARTH, F. (1956b): Indus and Swat Kohistan – an ethnographic survey. In: Studies honouring the Centennial of Universitetets Etnografiske Museum, Oslo 1857–1957; Vol.II. Oslo, 5–98.

BASHIR, E. & ISRAR-UD-DIN (eds.) (1996): Proceedings of the Second International Hindukush Cultural Conference. Karachi, Oxford, New York.

BEUERMANN, A. (1967): Fernweidewirtschaft in Südosteuropa. Braunschweig.

BHASIN, Veena (1996): Transhumants of Himalayas. Changpas of Ladakh, Gaddis of Himachal Pradesh and Bhutias of Sikkim. Delhi.

BISHOP, B. C. (1990): Karnali under stress. Livelihood strategies and seasonal rhythms in a changing Nepal Himalaya. Chicago (= Geography Research Paper 228–29).

BLACHE, J. (1934): L'homme et la montagne. Paris (= Géographie humaine 3).

BRUGGER, E. A. et al. (ed.) (1984): The transformation of Swiss mountain regions. Problems of development between self-reliance and dependency in an economic and ecological perspective. Bern, Stuttgart.

BRUSH, S. B. (1976a): Introduction, Symposium on Cultural Adaptation of Mountain Ecosystems. In: Human Ecology 4, 125–134.

BRUSH, S. B. (1976b): Man's Use of an Andean Ecosystem. In: Human Ecology 4 (2), 147–166.

BRUSH, S. B. & D. Guillett (1985): Small-scale agro-pastoral production in the Central Andes. In: Mountain Research and Development 5 (1), 19–30.

BUTZ, D. (1996): Sustaining indigenous communities: symbolic and instrumental dimensions of pastoral resource use in Shimshal, Northern Pakistan. In: The Canadian Geographer 40 (1), 36–53.

CASIMIR, M. J. & A. RAO (1985): Vertical Control in the western Himalaya: Some Notes on the Pastoral Ecology of the Nomadic Bakkarwal of Jammu and Kashmir. In: Mountain Research and Development 5 (3), 221–232.

CASIMIR, M. J. (1991): Pastoral strategies and balanced diets: two case studies from South Asia. In: BOHLE, H. G. et al. (eds.): Famine and food security in Africa and Asia. Indigenous responses and external intervention to avoid hunger. Bayreuth (= Bayreuther Geowissenschaftliche Arbeiten 15), 115–126.

CHAKRAVARTY-KAUL, M. (1998): Transhumance and customary pastoral rights in Himachal Pradesh: Claiming the high pastures for Gaddis. In: Mountain Research and Development 18 (1), 5–17.

CHAUBE, S. K. (ed.) (1985): The Himalayas: Profiles of Modernization and Adaptation. New Delhi.

CLEMENS, J. & M. NÜSSER (1997): Resource management in Rupal Valley, Northern Pakistan: The utilization of forests and pastures in the Nanga Parbat area. In: STELLRECHT, I. & M. WINIGER (eds.): Perspectives on history and change in the Karakorum, Hindukush, and Himalaya. Cologne (= Culture Area Karakorum Scientific Studies 3), 235–263.

CROOK, J. H. & H. A. OSMASTON (eds.) (1994): Himalayan Buddhist Villages. Environment, resources, society and religious life in Zanskar, Ladakh. Bristol.

EHLERS, E. (1976): Bauern – Hirten – Bergnomaden am Alvand Kuh/Westiran. In: UHLIG, H. & E. EHLERS (eds.): Tagungsbericht und wissenschaftliche Abhandlungen. 40. Deutscher Geographentag Innsbruck (19.–25.5.1975). Wiesbaden, 775–794.

EHLERS, E. (1980): Die Entnomadisierung iranischer Hochgebirge – Entwicklung und Verfall kulturgeographischer Höhengrenzen in vorderasiatischen Hochgebirgen. In: JENTSCH, C. & H. LIEDTKE (eds.): Höhengrenzen in Hochgebirgen (= Arbeiten aus dem Geographischen Institut des Saarlandes 29). Saarbrücken, 311–325.

EHLERS, E. (1995): Die Organisation von Raum und Zeit - Bevölkerungswachstum, Ressourcenmanagement und angepaßte Landnutzung im Bagrot/Karakorum. In: Petermanns Geographische Mitteilungen 139 (2), 105–120.

FORMAN, S. H. (1978): The future value of the "verticality" concept: implications and possible applications in the Andes. In: Actes du XLII Congrés International des Américanistes (Paris), 4, 233–256 [reprinted in: ALLAN, N. J. R., KNAPP, G. W. & C. STADEL (eds.) (1988): Human Impact on Mountains. Totowa, 133–153].

FRÖDIN, J. 1941: Zentraleuropas Alpwirtschaft. Oslo (= Instituttet for Sammenlignende Kulturforskning, Serie B 38, 2).

GOLDSTEIN, M. C. & C. BEALL (1990): Nomads of Western Tibet: The survival of a way of life. Berkeley.

GOLDSTEIN, M. C., BEALL, C. M. & R. P. CINCOTTA (1990): Traditional nomadic pastoralism and ecological conservation on Tibet's northern Plateau. In: National Geographic Research 6(2), 139–156.

GRÖTZBACH, E. (1972): Kulturgeographischer Wandel in Nordost-Afghanistan seit dem 19. Jahrhundert (= Afghanische Studien 4). Meisenheim am Glan.

GRÖTZBACH, E. (1976): Überlegungen zu einer vergleichenden Kulturgeographie altweltlicher Hochgebirge. In: Tagungsbericht und wissenschaftliche Abhandlungen. 40. Deutscher Geographentag Innsbruck (19.–25.5.1975). Wiesbaden, 109–120.

GRÖTZBACH, E. (1980): Die Nutzung der Hochweidestufe als Kriterium einer kulturgeographischen Typisierung von Hochgebirgen. In: JENTSCH, C. & H. LIEDTKE (eds.): Höhengrenzen in Hochgebirgen (= Arbeiten aus dem Geographischen Institut der Universität des Saarlandes 29). Saarbrücken, 265–277.

GRÖTZBACH, E. (1982): Das Hochgebirge als menschlicher Lebensraum. München (= Eichstätter

Hochschulreden 33) [English translation: High Mountains as Human Habitat. In ALLAN, N. J. R., KNAPP, G. W. & C. STADEL (eds.) (1988): Human Impact on Mountains. Totowa, 24–35].

GRÖTZBACH, E. (1984): Bagrot – Beharrung und Wandel einer peripheren Talschaft im Karakorum. In: Die Erde 115 (4), 305–321.

GRÖTZBACH, E. (1987): Zur siedlungsgeographischen Charakterisierung von Hochgebirgen. In: Frankfurter Beiträge zur Didaktik der Geographie 10, 61–75.

GRÖTZBACH, E. (1989): Kaghan – Zur Entwicklung einer peripheren Talschaft im Westhimalaya (Pakistan). In: Beiträge und Materialien zur regionalen Geographie 2, 1–18.

GRÖTZBACH, E. & C. STADEL (1997): Mountain peoples and cultures. In: MESSERLI, B. & J. IVES (eds.): Mountains of the World. A global priority. New York, London, 17–38.

GUILLET, D. (1983): Toward a Cultural Ecology of Mountains. The Central Andes and the Himalaya Compared. In: Current Anthropology 24/5, 561–574.

HAFFNER, W. (1985): Potentials and limits of agricultural production in Nepal as seen from an ecological-geographical standpoint. In: LAUER, W. (ed.): Natural environment and man in tropical mountain ecosystems. Stuttgart (= Erdwissenschaftliche Forschung 18), 115–126.

HERBERS, H. (1998): Arbeit und Ernährung in Yasin. Stuttgart (= Erdkundliches Wissen 123).

HERBERS, H. & G. STÖBER (1995): Bergbäuerliche Viehhaltung in Yasin (Northern Areas, Pakistan): organisatorische und rechtliche Aspekte der Hochweidenutzung. In: Petermanns Geographische Mitteilungen 139 (2), 87–104.

HEWITT, F. (1989): Woman's work and woman's place: The gendered life-world of a high mountain community in Northern Pakistan. In: Mountain Research and Development 9 (4), 335–352.

HEWITT, K. (1988): The study of mountain lands and peoples: a critical overview. In: ALLAN, N. J. R., KNAPP, G. W. & C. STADEL (eds.): Human Impact on Mountains. Totowa, 24–35.

HOFMEISTER, B. (1961): Wesen und Erscheinungsformen der Transhumance. Zur Diskussion um einen agrargeographischen Begriff. In: Erdkunde 15 (1), 121–135.

HONGLEI SUN (1985): Land Resource and Agriculture Utilization in Xizang Autonomous Region, China. In: SINGH, T. V. & J. KAUR (eds.): Integrated Mountain Development. New Delhi, 208–216.

HÜTTEROTH, W.-D. (1959): Bergnomaden und Yaylabauern im mittleren kurdischen Taurus. Marburg (= Marburger Geographische Schriften 11).

HÜTTEROTH, W.-D. (1973): Zum Kenntnisstand über Verbreitung und Typen von Bergnomadismus und Halbnomadismus in den Gebirgs- und Plateaulandschaften Südwestasiens. In: Rathjens, C., Troll, C. & H. Uhlig (eds.): Vergleichende Kulturgeographie der Hochgebirge des südlichen Asien (= Erdwissenschaftliche Forschungen 5). Wiesbaden, 146–156.

International Centre for Integrated Mountain Development (ICIMOD) (1998): Livestock development in mixed crop farming systems. Lessons and research priorities. Kathmandu, 4 pages.

ITURRIZAGA, L. (1997): The valley of Shimshal – a geographical portrait of a remote high settlement and its pastures with reference to environmental habitat conditions in the North-West Karakorum (Pakistan). In: GeoJournal 42 (2–3), 308–323.

IVES, J. D. & B. MESSERLI (1989): The Himalayan Dilemma. Reconciling Development and Conservation. London und New York.

JENTSCH, C. (1977): Für eine vergleichende Kulturgeographie der Hochgebirge. In: JENTSCH, C. (ed.): Beiträge zur geographischen Methode und Landeskunde. Festgabe für Gudrun Höhl. Mannheim, 57–71

JENTSCH, C. & H. LIEDTKE (eds.) (1980): Höhengrenzen in Hochgebirgen (= Arbeiten aus dem Geographischen Institut des Saarlandes 29). Saarbrücken.

JEST, C. (1973): La société pastorale du Tibet de lóuest. In: L'Homme, hier et aujourd' hui. Paris, 435–444.

JEST, C. (1974): Tarap, une vallée dans l'Himalaya. Paris.

JEST, C. (1975): Dolpo. Communautés de langue tibétaine du Népal. Paris.
JETTMAR, K. (1960): Soziale und wirtschaftliche Dynamik bei asiatischen Gebirgsbauern. In: Sociologus, N.S. 10, 120–138.
JODHA, N. S. (1997): Mountain agriculture. In: MESSERLI, B. & J. IVES (eds.): Mountains of the World. A global priority. New York, London, 313–335.
JODHA, N. S., BANSKOTA, M. & T. PARTAP (eds.) (1992): Sustainable mountain agriculture. New Delhi.
KARMYSCHEVA, B. C. (1969): Arten der Viehhaltung in den Südbezirken von Usbekistan und Tadshikistan. In: FÖLDES, L. (ed.): Viehwirtschaft und Hirtenkultur. Budapest, 112–126.
KARMYSCHEVA, B. C. (1981): Versuch einer Typologisierung der traditionellen Formen der Viehwirtschaft Mittelasiens und Kasachstans am Ende des 19./Anfang des 20. Jh. In: Die Nomaden in Geschichte und Gegenwart (= Veröffentlichungen des Museums für Völkerkunde zu Leipzig 33). Berlin, 91–96.
KHATANA, R. P. (1985). Gujara-Bakarwal transhumance in Jammu and Kashmir Himalayas: responses of spring migrations. In: HUSAIN, M, et al. (eds.): Geography of Jammu & Kashmir. Some Aspects. New Delhi, 86–116.
KREUTZMANN, H. (1986): A Note on Yak-keeping in Hunza (Northern Areas of Pakistan). In: Production Pastorale et Société 19, 99–106.
KREUTZMANN, H. (1989): Hunza – Ländliche Entwicklung im Karakorum. (= Abhandlungen – Anthropogeographie. Institut für Geographische Wissenschaften der Freien Universität Berlin 44). Berlin.
KREUTZMANN, H. (1991): The Karakoram Highway: The impact of road construction on mountain societies. In: Modern Asian Studies, 25 (4), 711–736.
KREUTZMANN, H. (1993a): Development trends in the high mountain regions of the Indian Subcontinent: a review. In: Applied Geography and Development 42, 39–59.
KREUTZMANN, H. (1993b): Challenge and Response in the Karakoram. Socio-economic transformation in Hunza, Northern Areas, Pakistan. In: Mountain Research and Development 13 (1), 19–39.
KREUTZMANN, H. (1994): Habitat conditions and settlement processes in the Hindukush-Karakoram. In: Petermanns Geographische Mitteilungen 138 (6), 337–356.
KREUTZMANN, H. (1995a): Mobile Viehwirtschaft der Kirgisen am Kara Köl. Wandlungsprozesse an der Höhengrenze der Ökumene im Ostpamir und im westlichen Kun Lun Shan. In: Petermanns Geographische Mitteilungen 139 (3), 159–178.
KREUTZMANN, H. (1995b): Globalization, spatial integration, and sustainable development in Northern Pakistan. In: Mountain Research and Development 15 (3), 213–227.
KREUTZMANN, H. (1995c): Communication and Cash Crop Production in the Karakorum: Exchange Relations under Transformation. In: STELLRECHT, I. (ed.): Pak-German Workshop: Problems of Comparative High Mountain Research with Regard to the Karakorum, Tübingen October 1992 (= Occasional Papers 2). Tübingen, 100–117.
KREUTZMANN, H. (1996): Ethnizität im Entwicklungsprozeß. Die Wakhi in Hochasien. Berlin.
KREUTZMANN, H. (1998a): From water towers of mankind to livelihood strategies of mountain dwellers: approaches and perspectives for high mountain research. In: Erdkunde 52 (3), 185–200.
KREUTZMANN, H. (1998b): Yak-keeping in High Asia. Adaptation to the utilization of high-altitude ecological zones and the socio-economic transformation of pastoral environments. In: Kailash (Katmandu) 18 (1–2), 17–38.
KREUTZMANN, H. (1999a): Rückzugsgebiet und Migration: Überlegungen zur Mobilität als Existenzsicherungsstrategie in Hochasien. In: JANZEN, J. (Hrsg.): Räumliche Mobilität und Existenzsicherung. Berlin (= Abhandlungen - Anthropogeographie 60), 83–104.
KREUTZMANN, H. (1999b): Animal Husbandry in High Asia. Yak-keeping at the upper pastoral limits. In: MIEHE, G. (ed.): Environmental Change in High Asia. Marburg (= Marburger Geographische Schriften; in print)

KRINGS, T. (1991): Indigenous agricultural development and strategies for coping with famine - the cases of Senoufo (Pomporo) in Southern Mali (West Africa). In: BOHLE, H. G. et al. (eds.): Famine and food security in Africa and Asia. Indigenous responses and external intervention to avoid hunger. Bayreuth (= Bayreuther Geowissenschaftliche Arbeiten 15), 69–81.
KUSSMAUL, F. (1965a): Badaxšan und seine Tağiken. Vorläufiger Bericht über Reisen und Arbeiten der Stuttgarter Badaxšan-Expedition 1962/63. In: Tribus 14, 11–99.
KUSSMAUL, F. (1965b): Siedlung und Gehöft bei den Tağiken in den Bergländern Afghanistans. In: Anthropos 60, 487–532.
LANGENDIJK, M. A. M. (1991): The Utilisation & Management of Pasture Resources in Central Ishkoman. Gilgit.
LAUER, W. (ed.) (1984): Natural environment and man in tropical mountain ecosystems. Stuttgart (= Erdwissenschaftliche Forschungen 18).
LAUER, W. (1987): Geoökologische Grundlagen andiner Agrarsysteme. In: Tübinger Geographische Studien 96, 51–71.
LAUER, W. & W. ERLENBACH (1987): Die tropischen Anden. Geoökologische Raumgliederung und ihre Bedeutung für den Menschen. In: Geographische Rundschau 39 (2), 86–95.
LICHTENBERGER, E. (1979): Die Sukzession von der Agrar- zur Freizeitgesellschaft in den Hochgebirgen Europas. In: Innsbrucker Geographische Studien 5, 401–436 [English translation: The Succession of an Agricultural Society to a Leisure Society: The High Mountains of Europe. In: ALLAN, N. J. R., KNAPP, G. W. & C. STADEL (eds.) (1988): Human Impact on Mountains. Totowa, 218–227].
MAHNKE, L. (1982): Zur indianischen Landwirtschaft im Siedlungsgebiet der Kallawaya (Bolivien). In: Erdkunde 36, 247–254.
MANZAR ZARIN, M. & R. L. SCHMIDT (1984): Discussions with Hariq. Land Tenure and Transhumance in Indus Kohistan. Islamabad und Berkeley.
MESSERLI, B. & J. IVES (eds.) (1997): Mountains of the World. A global priority. New York, London.
MILLS, M. (1996): Winds of Change: Women's traditional work and educational development in Pakora, Ishkoman Tehsil. In: BASHIR, E. & ISRAR-UD-DIN (eds.): Proceedings of the Second International Hindukush Cultural Conference. Karachi, Oxford, New York, 417–426.
MITCHELL, W. P. & D. GUILLET (ed.) (1994): Irrigation in High Altitudes: The Social Organization of Water Control Systems in the Andes. (= Society for Latin American Anthropology Publication Series 12).
MUKHIDDINOV, I. (1975): Zemledilie pamirskikh tadzhikov Vakhana i Ishkashima (Vol. XIX–nachale XX V.) (= Agricultural practices of the Pamir Tajik in Wakhan and Ishkashim). Moscow.
MURRA, J. V. 1972: El "Control Vertical" de un máximo de pisos ecológicos en la economía de las Sociedades Andinas. In: MURRA, J. V. (ed.): Visita de la provinncia de León de Huánuco en 1562. Tomo II. Huánuco, 427–468.
NAYYAR, A. (1986): Astor. Eine Ethnographie. Wiesbaden (= Beiträge zur Südasienforschung 88).
NÜSSER, M. & CLEMENS, J. (1996): Impacts on mixed mountain agriculture in Rupal Valley, Nanga Parbat, Northern Pakistan. In: Mountain Research and Development 16 (2), 117–133.
NÜSSER, M. (1998): Nanga Parbat (NW-Himalaya): Naturräumliche Ressourcenausstattung und humanökologische Gefügemuster der Landnutzung. Bonn (= Bonner Geographische Abhandlungen 97).
NÜSSER, M. (1999): Mobile Tierhaltung in Chitral: Hochweidenutzung und Existenzsicherung im pakistanischen Hindukusch. In: JANZEN, J. (Hrsg.): Räumliche Mobilität und Existenzsicherung. Berlin (= Abhandlungen - Anthropogeographie 60), 105–131.
ORLOVE, B. S. & D. W. GUILLET (1985): Theoretical and methodological considerations on the

study of mountain peoples: Reflections on the idea of subsistence type and the role of history in human ecology. In: Mountain Research and Development 5 (1), 3–18.

PANT, S. D. (1935): The social economy of Himalayas: based on a survey in the Kumaon Himalaya. London.

PARKES, P. (1987): Livestock symbolism and pastoral ideology among the Kafirs of the Hindu Kush. In: Man, N. S. 22, 637–660.

PEATTIE, R. (1936): Mountain Geography. A critique and field study. Cambridge, Mass.

PLANHOL, X. de (1968): Pression démographique et vie montagnarde particulièrement dans la ceinture alpino-himalayenne. In: Revue de Géographie Alpine 57 (3–4), 531–551.

PRICE, L. W. (1981): Mountains and Man. A Study of Process and Environment. Berkeley.

PRICE, M. F. & M. THOMPSON (1997): The complex life: human land uses in mountain ecosystems. In: Global Ecology and Biogeography Letters 6, 77–90.

PRICE, M. (ed.) (1999): Global change in the mountains. New York, London.

RAO, A. (1992): The constraints of nature or culture? Pastoral resources and territorial behaviour in the Western Himalayas. In: CASIMIR, M. J. & A. RAO (eds.): Mobility and Territoriality. New York, Oxford, 91-134.

RAO, A. & M. J. CASIMIR (1982): Mobile Pastoralists of Jammu and Kashmir. In: Nomadic Peoples 10, 40–50.

RAO, A. & M. J. CASIMIR (1985): Pastoral niches in the Western Himalayas (Jammu & Kashmir). In: Himalayan Research Bulletin 5 (2), 28–42.

RATHJENS, C. (1968): Neuere Entwicklungen und Aufgaben einer vergleichenden Geographie der Hochgebirge. In: Geographisches Taschenbuch 1966/69. Wiesbaden, 199–210 [reprint in: UHLIG, H. & W. HAFFNER (eds.) (1984): Zur Entwicklung der Vergleichenden Geographie der Hochgebirge (= Wege der Forschung 223). Darmstadt, 364–376].

RATHJENS, C. (1973): Fragen des Wanderhirtentums in vorder- und südasiatischen Hochgebirgsländern. In: Rathjens, C., Troll, C. & H. Uhlig (eds.): Vergleichende Kulturgeographie der Hochgebirge des südlichen Asien (= Erdwissenschaftliche Forschungen 5). Wiesbaden, 141–145.

RATHJENS, C. (1981): Terminologische und methodische Fragen der Hochgebirgsforschung. In: Geographische Zeitschrift 69, 68–77.

RATHJENS, C. (1982): Vergleich und Typenbildung in der Hochgebirgsforschung. In: Beiträge zur Hochgebirgsforschung und zur allgemeinen Geographie. Festschrift für Harald Uhlig Bd. 2 (= Erdkundliches Wissen 59). Wiesbaden, 1–8.

RATHJENS, C. (1988): German geographical research in the high mountains of the World. In: WIRTH, E. (ed.): German geographical research overseas. A report to the International Geographical Union. Tübingen.

RATHJENS, C., TROLL, C. & H. UHLIG (eds.) (1973): Vergleichende Kulturgeographie der Hochgebirge des südlichen Asien. Wiesbaden (= Erdwissenschaftliche Forschungen 5). Wiesbaden.

RHOADES, R. E. & S. I. THOMPSON (1975): Adaptive strategies in alpine environments: Beyond ecological particularism. In: American Ethnologist 2, 535–551.

RICHARDS, P. (1985): Indigenous agricultural revolution. London.

RIEDER, P. & J. WYDER (1997): Economic and political framework for sustainability of mountain areas. In: MESSERLI, B. & J. IVES (eds.): Mountains of the World. A global priority. New York, London, 85–102.

RINSCHEDE, G. 1979: Die Transhumance in den französischen Alpen und in den Pyrenäen. Münster (= Westfälische Geographische Studien 32).

RINSCHEDE, G. 1988: Transhumance in European and American Mountains. In: ALLAN, N. J. R., KNAPP, G. W. & C. STADEL (eds.): Human Impact on Mountains. Totowa, New Jersey, 96–108.

SCHICKHOFF, U. (1993): Das Kaghan-Tal im Westhimalaya (Pakistan). Bonn (= Bonner Geographische Abhandlungen 87).

SCHLEE, G. (1989): Identities on the move: clanship and pastoralism in Northern Kenya. London (= International African Library 5).

SCHMIDT-VOGT, D. (1990): High Altitude Forests in the Jugal Himal (Eastern Central Nepal). Stuttgart (= Geoecological Research 6).

SCHOLZ, F. (1982): Einführung. In: SCHOLZ, F. & J. JANZEN (eds.): Nomadismus - Ein Entwicklungsproblem? (= Abhandlungen des Geographischen Instituts - Anthropogeographie 33). Berlin, 2–8.

SCHOLZ, F. (ed.) (1991): Nomaden – Mobile Tierhaltung. Zur gegenwärtigen Lage von Nomaden und zu den Problemen und Chancen mobiler Tierhaltung. Berlin.

SCHOLZ, F. (1992): Nomadismus. Bibliographie. Berlin.

SCHOLZ, F. (1995): Nomadismus. Theorie und Wandel einer sozio-ökologischen Kulturweise. Stuttgart (= Erdkundliches Wissen 118).

SCHOLZ, F. (1995): Nomadismus ist tot. Mobile Tierhaltung als zeitgemäße Nutzungsform der kargen Weiden des Altweltlichen Trockengürtels. In: Geographische Rundschau 51 (5), 248–255.

SCHWEIZER, G. (1981): Die bergbäuerliche Wirtschaft des nordostanatolischen Randgebirges (Türkei) im Strukturwandel. In: Frankfurter Wirtschafts- und Sozialgeographische Schriften 36, 85–109.

SCHWEIZER, G. (1984): Zur Definition und Typisierung von Hochgebirgen aus der Sicht der Kulturgeographie. In: GRÖTZBACH, E. & G. RINSCHEDE (eds.): Beiträge zur vergleichenden Kulturgeographie der Hochgebirge (= Eichstätter Beiträge 12). Regensburg, 57–72.

SCHWEIZER, G. (1985): The changing structure of mountain farming in the Pontus Mountains (Eastern Black Sea Region, Turkey). In: SINGH, T. V. & J. KAUR (eds.): Integrated Mountain Development. New Delhi, 156–174.

SÈNE, EL HADJI & D. MCGUIRE (1997): Sustainable Mountain development - Chapter 13 in action. In: MESSERLI, B. & J. IVES (eds.): Mountains of the Worls. A global priority. New York, London, 447–453.

SHAHRANI, M. N. (1979): The Kirghiz and Wakhi of Afghanistan. Adaptation to Closed Frontiers. Seattle, London.

SHASHI, S. S. (1979): The Nomads of the Himalayas. Delhi.

SHEIKH, M. I. & S. M. KHAN (1982): Forestry and range management in Northern Areas. Peshawar.

SIEGER, R. (1907): Zur Geographie der zeitweise bewohnten Siedlungen in den Alpen. In: Verhandlungen des sechzehnten Deutschen Geographentages zu Nürnberg vom 21. bis 26. Mai 1907, 262–272 [reprinted as well in: Geographische Zeitschrift 13 (1907), 361–369].

SINGH, T. V. & J. KAUR (eds.) (1985): Integrated Mountain Development. New Delhi.

SKELDON, R. (1985): Population pressure, mobility, and socio-economic change in mountainous environments: regions of refuge in comparative perspective. In: Mountain Research and Development 5 (3), 233–250.

SNOY, P. (1965): Nuristan und Munǧan. In: Tribus 14, 101–148.

SNOY, P.(1993): Alpwirtschaft in Hindukusch und Karakorum. In: SCHWEINFURTH, U. (eds.): Neue Forschungen im Himalaya (= Erdkundliches Wissen 112). Stuttgart, 49–73.

SOFFER, A. (1982): Mountain geography - a New Approach. In: Mountain Research and Development 2, 391–398.

STADELBAUER, J. (1984): Bergnomaden und Yaylabauern in Kaukasien. Zur demographischen Entwicklung und zum sozioökonomischen Wandel bei ethnischen Gruppen mit nichtstationärer Tierhaltung. In: Paideuma 30, 201–229.

STALEY, E. (1966): Arid Mountain Agriculture in Northern West Pakistan. Lahore.

STALEY, J. (1966): Economy and Society in Dardistan: Traditional Systems and the Impact of Change. Lahore.

STALEY, J. (1969): Economy and Society in the High Mountains of Northern Pakistan. In: Modern Asian Studies 3, 225–243.

STALEY, J. (1982): Words for my brother. Travels between the Hindu Kush and the Himalayas. Karachi.
STELLRECHT, J. (ed.) 1998: Karakorum – Hindukush – Himalaya: Dynamics of Change. Cologne (= Culture Area Karakorum Scientific Studies 4).
STELLRECHT, I. & M. WINIGER (eds.) (1997): Perspectives on history and change in the Karakorum, Hindukush, and Himalaya. Cologne (= Culture Area Karakorum Scientific Studies 3).
STEVENS, S. F. (1993): Claiming the High Ground. Sherpas, subsistence, and environmental change in the Highest Himalaya. Berkeley, Los Angeles, London.
STÖBER, G. (1978): Die Afshar – Nomadismus im Raum Kerman (Zentraliran). Marburg (= Marburger Geographische Schriften 76).
STÖBER, G. (1979): "Nomadismus" als Kategorie. In: EHLERS, E. (ed.): Beiträge zur Kulturgeographie des islamischen Orients (= Marburger Geographische Schriften 78). Marburg, 11–24.
STÖBER, G. (1988): The influence of politics on the formation and reduction of "ethnic boundaries" of tribal groups: the cases of Sayâd and Afsâr in eastern Iran. In: DIGARD, J.-P. (ed.): Le fait ethnique en Iran et en Afghanistan. Paris, 131–138.
STONE, P. B. (ed.) (1992): The State of the World's Mountains. A Global Report. London, New Jersey.
STRAND, R. (1975): The changing herding economy of the Kom Nuristani. In: Afghanistan Journal 2 (4), 123–134.
STREEFLAND, P. H., KHAN, S. H. & O.VAN LIESHOUT (1995): A contextual study of the Northern Areas and Chitral. Gilgit.
THOMPSON, M., WARBURTON, M. &. T. HATLEY (1986): Uncertainty on a Himalayan Scale. London.
TROLL, C. (1941): Studien zur vergleichenden Geographie der Hochgebirge der Erde. Bonn (= Bonner Mitteilungen 21).
TROLL, C. (1962): Die dreidimensionale Landschaftsgliederung der Erde. In: LEIDLMAIR, A. (ed.): Hermann von Wissmann-Festschrift. Tübingen, 54–80.
TROLL, C. (1972): Geoecology and the world-wide differentiation of high-mountain ecosystems. In: TROLL, C. (ed.): Geoecology of the high-mountain regions of Eurasia. Wiesbaden (= Erdwissenschaftliche Forschungen 4), 1–16.
TROLL, C. (1973): Die Höhenstaffelung des Bauern- und Wanderhirtentums im Nanga Parbat-Gebiet (Indus-Himalaya). In: Erdwissenschaftliche Forschungen 5, 43–48.
TROLL, C. (1975): Vergleichende Geographie der Hochgebirge in landschaftsökologischer Sicht. In: Geographische Rundschau 27, 185-198 [English translation: Comparative Geography of High Mountains of the World in the View of Landscape Ecology: A Development of Three and a Half Decades of Research and Organization. In: ALLAN, N. J. R., KNAPP, G. W. & C. STADEL (eds.) (1988): Human Impact on Mountains. Totowa, 36–56].
TUCKER, R. P. (1986): The evolution of transhumant grazing in the Punjab Himalaya. In: Mountain Research and Development 6 (1), 17–28.
UHLIG, H. (1962a): Kaschmir. In: Geographisches Taschenbuch 1962/63. Wiesbaden, 179–196.
UHLIG, H. (1962b): Typen der Bergbauern und Wanderhirten in Kaschmir und Jaunsar-Bawar (= Tagungsbericht und wissenschaftliche Abhandlungen. Deutscher Geographentag Köln 1961). Wiesbaden, 212–222.
UHLIG, H. (1965): Die geographischen Grundlagen der Weidewirtschaft in den Trockengebieten der Tropen und Subtropen (mit Beispielen aus Kaschmir und Kolumbien). In: Gießener Beiträge zur Entwicklungsforschung 1, 1–28.
UHLIG, H. (1970): Die Agrarlandschaften des Chenab-Tales in Jammu und Kaschmir. In: Beiträge zur Geographie der Tropen und Subtropen (= Tübinger Geographische Studien 34). Tübingen, 309–323.
UHLIG, H. (1973): Wanderhirten im westlichen Himalaya: Chopan, Gujars, Bakerwals, Gaddi. In: RATHJENS, C., TROLL, C. & H. UHLIG (eds.): Vergleichende Kulturgeographie der Hochgebirge des südlichen Asien (= Erwissenschaftliche Forschungen 5). Wiesbaden, 157–167.

UHLIG, H. (1976): Bergbauern und Hirten im Himalaya. In: UHLIG, H. & E. EHLERS (eds.): Tagungsbericht und wissenschaftliche Abhandlungen. 40. Deutscher Geographentag Innsbruck (19.–25.5.1975). Wiesbaden, 549–586.

UHLIG, H. (1978): Geoecological controls on High-Altitude Rice Cultivation in the Himalayas and Mountain Regions of Southeast Asia. In: Arctic and Alpine Research 10, 519–529.

UHLIG, H. (1980): Der Anbau an den Höhengrenzen der Gebirge Süd- und Südostasiens. In: JENTSCH, C. & H. LIEDTKE (eds.): Höhengrenzen in Hochgebirgen (= Arbeiten aus dem Geographischen Institut des Saarlandes 29). Saarbrücken, 279–310.

UHLIG, H. (1984): Die Darstellung von Geo-Ökosystemen in Profilen und Diagrammen als Mittel der Vergleichenden Geographie der Hochgebirge. In: GRÖTZBACH, E. & G. RINSCHEDE (eds.): Beiträge zur vergleichenden Kulturgeographie der Hochgebirge (= Eichstätter Beiträge 12). Regensburg, 93–152.

UHLIG, H. (1995): Persistence and Change in High Mountain Agricultural Systems. In: Mountain Research and Development 15 (3), 199–212.

Uhlig, H. & E. Ehlers (eds.) (1976): Tagungsbericht und wissenschaftliche Abhandlungen. 40. Deutscher Geographentag Innsbruck (19.–25.5.1975). Wiesbaden 1976.

UHLIG, H. & W. HAFFNER (eds.) (1984): Zur Entwicklung der Vergleichenden Geographie der Hochgebirge. Darmstadt (= Wege der Forschung 223).

VEYRET, P. & G. VEYRET (1962): Essai de définition de la monatgne. In: Revue de Géographie Alpine 50 (1), 5–35.

WILSON, R. T. (1997): Livestock, pastures, and the environment in the Kyrgyz Republic, Central Asia. In: Mountain Research and Development 17 (1), 57–68.

WINIGER, M. (1996): Karakorum im Wandel – Ein metodischer Beitrag zur Erfassung der Landschaftsdynamik in Hochgebirgen. In: Festschrift Bruno Messerli, Jb. Geogr. Ges. Bern, Vol. 59: 59–74.

# ANIMAL HUSBANDRY IN DOMESTIC ECONOMIES: ORGANIZATION, LEGAL ASPECTS AND PRESENT CHANGES OF COMBINED MOUNTAIN AGRICULTURE IN YASIN (NORTHERN AREAS, PAKISTAN)

Georg Stöber & Hiltrud Herbers

## 1. INTRODUCTION

Domestic economies are conceived as systems in which production and reproduction are organized on the household level and which do not differentiate between "business" and the "home". They mainly aim to meet the needs of the individual rather than to maximize profits. This is achieved co-operatively by all the members of a household, who also share the yields of their economic activities. However, household economies are not necessarily subsistence economies, nor are the home-produced goods exclusively consumed by the household members. On a small scale, both the production of commodities for sale and the additional purchase of items to supplement their own supply are possible. This does not change the principal aim of household economies, i.e. to meet the needs of the household members.

Domestic economies contrast with systems in which production and reproduction represent separate aims. This is the case, for instance, in industrial enterprises and the private households of their workers and employees. Whereas the production of enterprises is intended to maximize profit, private households are concerned with reproduction only. Keeping aside typological differences between the domestic and capitalist form of economies, the domestic mode is not uniform but domestic economies differ greatly according to the specific activities, which the people pursue in order to earn a living (e.g. agriculture or handicraft), and to the local ecological and socio-political conditions.[1] This paper discusses major aspects of a form of domestic economy, which is known as mixed or combined mountain agriculture, and which is typical for many parts of the Hindu-kush, the Karakoram, and other high mountain areas.

Combined mountain agriculture can briefly be defined as an interdependent combination of crop cultivation and animal husbandry that makes use of different ecological zones.[2] This practice implies an elaborated management of all associ-

---

1  Cf. the case studies in Durrenberger (1984). The concept of domestic economy and its theoretical implications discussed by Tschajanow (1923) have been the target of much academic debate and criticism.
2  Both aspects have been dealt with in a number of articles. Cf. Haserodt (1989) for Chitral, Grötzbach (1984) for Bagrot, Kreutzmann (1989) for Hunza, and Clemens & Nüsser

ated activities. This paper intends to describe the organization of combined mountain agriculture in the Yasin Valley in Northern Pakistan. Due to the context of this volume we will pay particular attention to animal husbandry and the use of alpine pastures. Thus, we will point out the legal preconditions of the use of high mountain areas and discuss the contemporary changes. This discussion, too, however, will demonstrate that grazing is an integrated part of the economy, understandable only when not seen in isolation.

## 2. THE YASIN VALLEY

Yasin is located in the Northern Areas of Pakistan.[3] Apart from the main valley, it includes several side-valleys, such as Nazbar, Thuibar, Qorkultibar, and Asumbar. The bottom of the valley rises from 2,160 m in the south up to over 2,700 m near the village of Darkot. Especially north of Darkot, where a lofty range separates the catchment area of the Yasin from that of the Yarkhun River, the surrounding peaks reach altitudes of more than 6,000 m.

Villages and fields are located along the river, mostly on alluvial fans of tributaries. Due to the arid conditions at the valley bottom and at the lower slopes,[4] agriculture is only possible by means of irrigation. The streams and torrents from which irrigation channels are diverted originate from glaciers, which are extensive especially in the northern part of the area (cf. STÖBER 2000).

The gradual increase in precipitation and the concomitant decrease in temperature from the valley bottom to higher altitudes[5] are the main constituents in the development of the several vegetation zones:

- Those parts of the riverbed, which are less affected by floods, are covered with so-called "forest" (*mushk*), which mainly consist of tamarisks.
- In the semi-desert of the valley bottom and adjacent slopes the people have created oases, i.e. irrigated fields to grow food and fodder crops.
- Altitudes between 2,800 m and 3,700 m are dominated by artemisia steppes (MIEHE et al. 1994). In this zone, summer settlements are situated with fields at the lower and without cultivation at higher altitudes.
- Further uphill, there follows a belt characterized by alpine grassland on the slopes and thick patches of grass and herbs in moist depressions.
- Between 4,200 m and 4,400 m the vegetation is replaced by detritus, snow and ice, with glacier tongues often reaching down to much lower altitudes.

(1994) for Astor. SNOY (1993) gives a general overview on the use of high pastures in the Hindu Kush and Karakoram, which includes anthropological aspects.

3   The Northern Areas belong to the territory disputed between India and Pakistan since 1947. Formerly known as the Gilgit Agency, the Northern Areas are – beside Azad Kashmir – the part of the territory disputed that at present is under the administration of Pakistan.
4   Measurements of precipitation carried out in the valley revealed average values which were below 150 mm per year (cf. WHITEMAN 1985: fig. 1).
5   According to OWEN and DERBYSHIRE (1989:34) the annual precipitation amounts to 2,000 mm at altitudes above 4,500 m. For the central parts of the Karakoram, HEWITT (1989:14) calculates a precipitation of 1,000–1,800 mm for altitudes between 5,000 m and 7,000 m.

These vegetation zones vary considerably in terms of their value as pasture. The alpine grasslands possess the highest value, even if they can only be used during the summer season when they are not covered with snow. By contrast, the artemisia steppes are free of snow for a longer period, but of lower fodder value. The wastelands at lower altitudes hardly lend themselves to regular use as pastures, while the irrigated areas provide several sources of fodder (e.g. weeds from the fields, grass from the fringes of the fields and small plots of irrigated meadows). Goats and sheep sometimes also graze in the *mushk* of the riverbed, an area which is easily accessible during the winter season although it is only of very limited use as a source of forage.

The population of Yasin amounts to about 30,000 inhabitants. They are divided into various patrilineal descent groups (*qóom*) of distinct origin and social prestige. The larger villages generally consist of a number of hamlets which are mostly inhabited by members of one *qóom*. Households of these hamlets often combine into larger groups (*giram* or *mon*) in order to co-operate in socio-religious and agricultural activities, such as entertaining guests at weddings and funerals, or spreading manure in the fields, respectively.

## 3. COMBINED MOUNTAIN AGRICULTURE IN YASIN

For most of the households, agriculture plays an important role, even though nowadays it is not the exclusive means to secure food supplies. People in Yasin usually combine the cultivation of crops with animal husbandry. Since most of the households possess only small plots of arable land, production mainly serves to meet the needs of the household. Even if one takes into account that a number of families own land in several villages – which they either cultivate themselves or rent out – and that many farmers possess land in their summer settlements,[6] there is only a small percentage of households which dispose of more than one hectare of arable land (cf. Tab. 1). In many cases, the household's landholding is even much smaller. Since every son is entitled to an equal share of the household's fields in the case the household splits after the death of the father, the plots become increasingly fragmented.

The main crop in the villages is generally wheat. The cultivation of other cereals – maize, barley, and millet (*Panicum*) – vary with the size of the landholding. Thus, if a household has only some small fields it will cultivate mainly wheat and maize, or alternatively wheat and barley. The flour of these cereals is used to prepare bread, which is the basic food of the local people. If more land is available, many farmers cultivate additional crops, such as millet and pulses, or they extend the proportion of alfalfa.[7] Moreover, various vegetables and fruits are grown in small gardens.

---

6  In this paper, we distinguish between summer settlements and alpine settlements. Other than the villages, both are used seasonally, but while the former is utilized for the cultivation of crops and animal husbandry, the latter is only used to keep livestock.

7  Alfalfa is generally sown on new fields, which are cultivated for the first time. It remains

Tab. 1a: Size of holdings (in *kanal**) in Yasin, 1982/83

| Village | Number of farmsteads | Per cent of households with | | | |
| --- | --- | --- | --- | --- | --- |
| | | <20 kanal | >20 to <40 kanal | >40 to <60 kanal | >60 kanal |
| Qorkulti | 71 | 54 | 39 | 6 | 1 |
| Thui | 490 | 50 | 40 | 6 | 4 |
| Yasin | 304 | 42 | 35 | 16 | 7 |
| Sandi | 226 | 38 | 43 | 9 | 10 |
| Barkulti | 274 | 36 | 38 | 18 | 8 |
| Taus | 210 | 25 | 54 | 12 | 9 |

Source: SAUNDERS (1983:172–204)

Tab. 1b: Holdings of farmers living in Sultanabad (Yasin), 1991

| Location of holdings | Per cent of households with holdings (in *kanal*) | | | | | |
| --- | --- | --- | --- | --- | --- | --- |
| | without land | ≤20 | >20 to ≤40 | >40 to ≤60 | >60 | un-tilled |
| Sultanabad | 2 | 46 | 38 | 12 | 2 | – |
| Summer settlements | 60 | 34 | 4 | – | – | 2 |
| Other main villages | 84 | 10 | 4 | 2 | – | – |
| All holdings | – | 28 | 40 | 28 | 4 | – |

Source: sample (n=50), G. Stöber 1991
* As no land surveying has taken place in Yasin, a *kanal* is defined locally as the area at which 10 kg of wheat seeds are used. This area varies between 100–300m$^2$.

Other factors, which determine the distribution of crops, are altitude and local climatic conditions. Southern Yasin is located in a transitional zone between the double and the single cropping area in which two grain crops can be harvested a year if maize is cultivated after barley. In areas, which allow only one harvest, barley instead of wheat is grown if soil is of poor quality. However, since wheat, which the people prefer as food grain, is nowadays available on the local markets, the cultivation of barley decreases. In Central Yasin it virtually disappeared several years ago. In the villages there, generally two thirds or three quarters of the fields are cultivated with wheat, the rest with maize. At higher altitudes, maize yields decrease. Thus, this crop is often replaced by barley.

In some side valleys, wheat is also grown beyond the villages even in the lower summer settlements located at altitudes up to approximately 3,000 m. However, the main cereal grown in the summer settlements is barley, which can be found up to altitudes of 3,500 m. Above this altitude, only alpine settlements are established, which are not used for cultivating crops.

> there for nearly ten years, before cereals can be sown. Due to the scarcity of fodder, alfalfa is a profitable cash crop. It is sold to other farmers and to Gujur, goat herders who immigrated to Yasin decades ago.

# Animal Husbandry in Domestic Economies

Photo 1
A village herd climbs up the slopes above Sultanabad village (central Yasin) on its morning drive to the daily pastures. Implementing the *nabat* practice, the herd is attended by one of the members of the herding group. (Photo: G. Stöber, June 1990)

Photo 2
Most of the home-produced goods are used to meet the needs of the household members. However, on a small scale these products are sold on the local markets. This woman from Yasin, for instance, prepared the dried milk product *qurút* for sale. (Photo: H. Herbers, September 1993)

In the permanent settlements, orchards and gardens provide additional food, which supplement the products from the fields. Fruit trees include apples, pears, walnuts, almonds, peaches and mulberries and to a lesser extent, cherries, plums, yellow plums, pomegranates, and sometimes vine. The major fruit trees, however, are numerous varieties of apricots. Apricots figure not only prominently in the local diets; dried apricots and apricot seeds are also important goods for sale.[8] Except potatoes, vegetables are not sold in Yasin. Thus, carrots, cabbage, onions, tomatoes, turnips, pumpkins, and various greens (e.g. *Malva sylvestris*) are consumed exclusively by the household members.

Apart from cultivating crops, farmers in Yasin keep goats, sheep, cattle, donkeys, sometimes yaks as well as poultry. A survey, carried out by Georg Stöber in Sultanabad in 1991, revealed only a third of the farmers as having two oxen, which are necessary to plough a field. Thus, most farmers have to borrow one ox or even both animals for their fieldwork.[9] Almost every household has at least one cow to provide its members with milk; according to the survey, three quarters of the households interviewed had one to three cows. In poorer households which cannot afford a cow, milk is provided by one or two goats. But the absolute number of goats – and to a lesser extent sheep – per household varies considerably. While more than half of the households in Sultanabad owned not more than ten sheep and goats, there were some farmers who had over forty. In the higher villages the average number of cows per household decreases, whereas that of sheep and goats increases (see Tab. 2). While the number of donkeys roughly equals that of oxen, horses are rare among the peasants and are mostly owned by polo players. The semi-domesticated yaks, which are exclusively used to provide meat and wool, are only kept in peripheral parts of the valley, especially in Thuibar, Nazbar, and in Northern Yasin.

The principal aim of domestic economies is the reproduction of the household members. In its comprehensive application, the term 'reproduction' covers a wide range of aspects, which serve to meet both the material/physical and immaterial/social needs of the people. On the one hand, there is the maintenance of the individual's living and working capacity (cf. KITSCHELT 1987:13). Activities that serve these ends include not only basic activities, such as eating, sleeping and recreating, but also comprise the preparation of meals, the provision of clothes and accommodation, health care etc. On the other hand, the regeneration of the individual is to be supplemented by generative reproduction. Children have to be born and brought up if the household is to survive one generation. Last but

---

8   Beside fruit trees, the farmers cultivate poplars, willows, and *Elaeagnus parvifolia* along the irrigation channels in order to provide timber.
9   Oxen are lent to other households for a period of two to three days either on a reciprocal or unilateral basis. The latter is the case especially if both animals have to be borrowed. The farmer, who borrows the oxen, does not have to pay a fee, but he is expected to feed the animals properly. Another widespread system in Yasin to exchange animals is called *suni*: Female animals can be borrowed from wealthy relatives for a period of several months in order to extend the household's livestock size. This means, female offspring of the borrowed animal remain in the household, while male offspring have to be returned to the owner.

Tab. 2: Number of animals in Yasin villages, 1982/83

| Animals | Yasin Village | Taus | Sandi | Bar-kulti | Thui | Qor-kulti | Darkot (1990)* |
|---|---|---|---|---|---|---|---|
| Goats | 2,190 | 1,453 | 1,261 | 1,887 | 3,220 | 551 | 1,343 |
| Sheep | 644 | 573 | 624 | 846 | 2,538 | 388 | 549 |
| Sheep and goats | 2,834 | 2,026 | 1,885 | 2,733 | 5,758 | 939 | 1,892 |
| Cattle | 1,409 | 1,006 | 965 | 1,470 | 1,669 | 214 | 501 |
| Yaks | – | – | – | 55 | 156 | 8 | (–) |
| Donkeys and horses | 335 | 270 | 270 | 316 | 552 | 75 | 168 |
| Poultry | 1,884 | 1,375 | 1,189 | 1,183 | 1,410 | 430 | 1,097 |
| Number of farms | 304 | 210 | 226 | 274 | 490 | 71 | 150 |
| Animals per household: | | | | | | | |
| Sheep and goats | 9.3 | 9.6 | 8.3 | 10.0 | 11.8 | 13.2 | 12.6 |
| Cattle | 4.6 | 4.8 | 4.3 | 5.4 | 3.4 | 3.0 | 3.3 |
| Yaks | – | – | – | 0.2 | 0.3 | 0.1 | – |
| Donkeys/horses | 1.1 | 1.3 | 1.2 | 1.2 | 1.1 | 1.1 | 1.1 |
| Poultry | 6.2 | 6.5 | 5.3 | 4.3 | 2.9 | 6.1 | 7.3 |

Sources: after SAUNDERS 1983: Appendix 4; *= unpublished data of AKRSP, Gilgit

not least, social relations have to be maintained, reproduced, as the individuals – and the households – are part of a wider social network which among others is a precondition of the afore-mentioned aspects of reproduction. The production of goods in a domestic economy, which is discussed in this paper, serves all those aspects of the term 'reproduction' as it enables the individuals to keep themselves alive, to propagate themselves and to practise social interaction.

Figure 1 demonstrates connections between the productive and reproductive aspects of Yasin's domestic economy in a rather general manner. The flow of labour (energy) and goods (matter) link not only productive and reproductive spheres, but also various elements belonging to the sphere of production. Some examples will illustrate this statement:

- On the one hand, oxen and donkeys are used for ploughing, threshing and carrying loads. Moreover, the manure of the livestock improves the soil fertility of the fields.[10] On the other hand, residues (e.g. straw, weeds, plant roots) and alfalfa from the fields are used to feed the animals in winter.[11]
- Cereals, vegetables, and fruits are mainly grown for subsistence. Various plant products are dried and stored to be consumed during the winter season. However, since most households do not produce enough grain to meet their nutritional needs (and have become accustomed to goods that are not pro-

---

10  In the villages, the manure of the cattle is spread on the fields. By contrast, manure of sheep and goats, which is regarded as particular fertile, is only used for vegetable gardens. In the summer settlements, however, the manure of various species is not separated in order to use it for different purposes.

11  The various species of animals differ in their fodder requirements. While hay is given to cattle, sheep and goats, straw and alfalfa is fed only to the former.

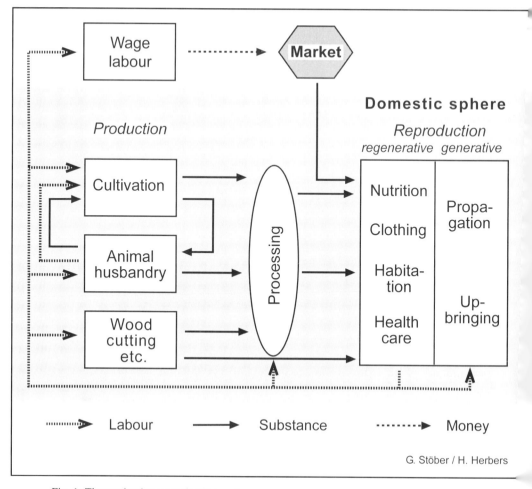

Fig. 1: The production-reproduction cycle

duced locally), they have to purchase additional food from the markets. This is mostly done with off-farm income of one or the other member of the household who has found employment in Yasin or elsewhere.
– The livestock provides the households also with wool, hair, and hides. These raw materials can be used to make pullovers, socks, carpets, shoes, leather sacks etc. Furthermore, various foods resulting from animal husbandry enrich the local diet. Milk is either used to prepare tea or to produce butter, *qurút*, and other dairy products. Moreover, almost every family slaughters an animal at the end of the year. Thus, people consume a modest amount of meat during the winter season.[12] At the same time, the custom to slaughter animals in winter helps to reduce the need for limited fodder resources.

---

[12] For further information on dietary and nutritional aspects of animal husbandry in high mountain areas cf. HERBERS in this volume.

- In order to cook meals, to process plant and animal raw material, or to heat the houses, the farmers of Yasin cut juniper trees on the high pastures and collect *Artemisia* on the slopes.
- A crucial prerequisite for the maintenance of the entire production-reproduction-cycle is the availability of male and female labour. Its regeneration depends on food, shelter, clothing, health care etc. Only when these means are sufficiently supplied, people are able to pursue the many tasks associated with cultivation, animal husbandry, food processing and other productive and reproductive activities (HERBERS 1998).

## 4. SPATIAL AND TEMPORAL ORGANIZATION OF AGRICULTURAL ACTIVITIES

The cultivation of crops and livestock breeding are not only linked in terms of material interdependencies. Both sectors have to be adjusted one to the other in spatial and temporal terms: the manifold agricultural tasks and the use of pastures at different altitudes have to be co-ordinated.

In the central part of Yasin, the agricultural season starts at the beginning of March. The farmers first mend the walls and fences made of dry thorn bushes. While the fields are freely accessible during the winter season, they are now closed off in order to prevent the livestock from damaging the crops. Further activities pursued in early spring include carrying the manure from the stables to the fields, spreading manure on the fields, and cleaning the irrigation channels, which are usually put into service in April.

Farmers commence sowing between the first half of March and April depending on the altitude of the settlements and the crops grown. Generally, barley is the first crop sown, but as in the central part of the Yasin Valley the farmers ceased to grow barley, wheat is the first crop there. After barley and wheat follow maize and – in May or June – millet. Farmers who grow barley in the summer settlements do the sowing during the first half of May, when the snow has melted away, i.e. before the sowing of the last crops has been completed in the villages. During the growth period all the fields have to be irrigated regularly.

Harvest time begins with the harvest of barley in the middle of June in the south; in the highest villages it starts during the first half of August. In the summer settlements, barley has to be reaped from mid-August to the beginning of September, i.e. at a time when the wheat harvest is not yet completed in the villages. Maize and millet harvest will follow so that harvest time is generally finished by the end of September (cf. Fig. 2).

Cultivating cereals at different altitudes adds to the workload of the people, because they have to walk long distances between the various locations of agricultural activities. On the other hand, however, this practice extends the periods in which sowing and harvesting have to be done. This levels off the peaks of the workload to some extent compared to an equivalent cultivation of one crop at one location.

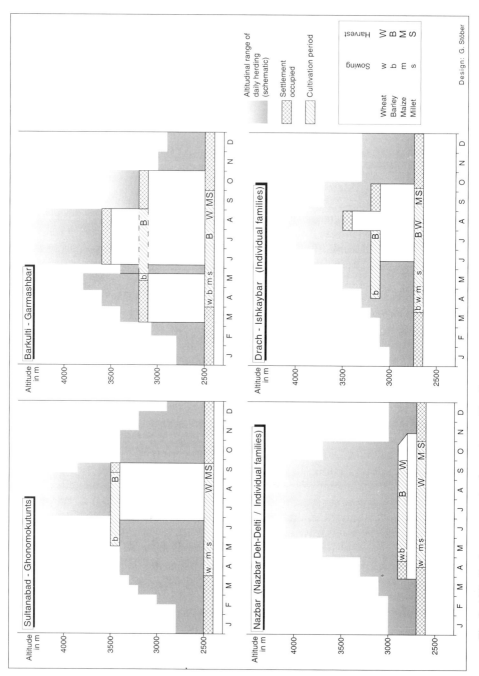

Fig. 2: Altitudinal arrangements of cropping and herding

As already mentioned, animal husbandry is well-attuned to the cultivation of crops. After sowing in early spring, the walls and fences around the fields are closed off to prevent the animals from entering the plots and destroying the shoots. During this time, sheep and goats are driven up daily to the pastures located on the slopes in the vicinity of the villages. Only cattle remain near the homestead to be grazed on wasteland or along the channels. Except one dairy cow or goat, the whole livestock is taken to the high pastures for the summer months. Villagers with pasture grounds at intermediate altitudes sometimes take the animals there as early as March or April. The pastures at higher altitudes, however, are visited at the end of June or July, when the snow has melted. The livestock returns from the high pastures to the villages in September or October. By this time, the harvest has been completed in the valley. Thus, the animals can graze on the stubble-fields. Sheep and goats, however, are generally taken to the same pastures as in spring until snowfall starts. During the winter, the livestock has to be fed. Only the *mushk* in the riverbed provides some additional fodder resources during this season.

Unlike the other animals, yaks graze unattended at higher altitudes all year round. It is only during times of heavy snowfall that they approach the villages where they are temporarily fed by the farmers.

The general patterns of the utilization of the pastures explained above are subject to local modifications. For example, the winter and spring pastures of several villages differ. At some places, farmers temporarily use the slopes on the other side of the valley, opposite their own village. This is common practice, for instance, in Harp between December and March, in Manich between September and March, and in Haltar during times of heavy snowfall.[13]

In addition, on some of the high pastures farmers use several grazing grounds at different altitudes successively, often in order to separate crops and livestock. The following examples will illustrate this procedure:

– In Barkulti, part of the population owns summer fields and grazing land in Garmashbar (cf. Figs. 2 and 3). In 1991, about 80 families took their livestock to these summer settlements located at an altitude of between 2,900 m and 3,300 m in March. After tilling the fields in May, they returned to Barkulti in order to let the seedlings grow undisturbed. Around 10th June, ten of these households took all the animals up to a pasture ground at 3,600 m. During this time, the summer fields were irrigated and later, in August, harvested by those who had remained in Barkulti. At the beginning of September, the animals were driven back from the alpine pastures to the stubble-fields in the summer settlements, and by the end of this month, or some time later, the households finally returned to Barkulti. Thus, people in Garmashbar separate the grazing grounds from their summer fields vertically.

– A different pattern is to be found in Nazbar, Asumbar and some of the summer settlements in Qorkultibar or Dasbar. In these areas, farmers usually

---

13 In Haltar these slopes are said to receive less snow than the 'normal' pastures and are therefore accessible even in wintertime.

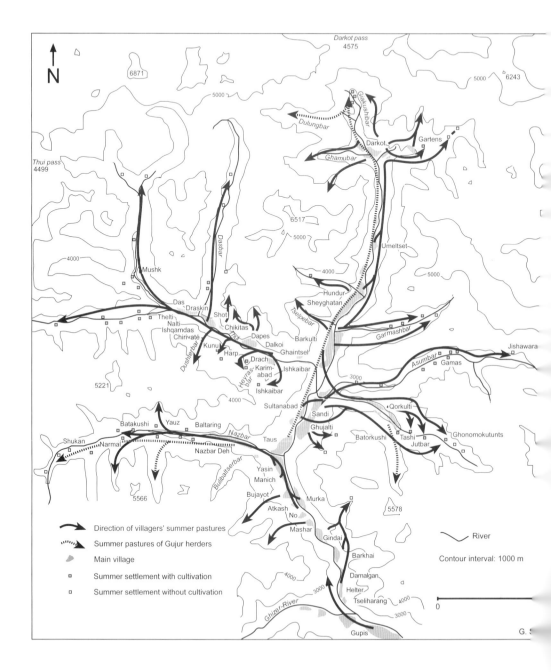

Fig. 3: High pastures in Yasin

own fields and lodgings on both banks of the river, which are used alternately: while fields are cultivated on one bank, the household and livestock use the lodgings on the other, and vice versa.[14] Although this system puts those households at a disadvantage which have property on one side only, the families affected still adhere to it since deviation would endanger the crop. However, households, which own land only in one location, often have the opportunity to lease additional plots of land.

- Occasionally, sites in different valleys are used alternately in two-year intervals. So farmers in Shot (Thui Valley) cultivate plots of land in Dasbar and Mushibar successively and send their flocks to the other place. After the harvest, however, the livestock is brought in to pasture on the stubble-fields for some weeks so that within the course of a year both locations are used, notwithstanding the afore-mentioned alternation.
- While the livestock is usually driven to the stubble-fields immediately after the harvest, sometimes a short stay at higher altitudes will proceed. For instance, some families in Ishkaybar move to higher settlements for a period of three weeks before returning to their summer village.[15]
- A special case is to be found in Giakushibar, one of the alpine pastures of Darkot village, in so far as it is used by two distinct groups, by households from Darkot itself and by families of the Gujur, who are specialized goat herders and use the Giakushibar pastures to graze their livestock. In spring, the Gujur move from Taus, having spent the winter there, to Giakushibar. Here they stay until summer. Then they move further up to Dulungbar, a second pasture located at a higher altitude (cf. Fig. 3), to vacate the place for its original owners, and the families from Darkot occupy their summer abodes in Giakushkibar where they cultivate barley. After their return to Darkot at the end of October, the Gujur come down again to spend the rest of the year in Giakushibar until they are forced to return to Taus with the onset of winter snows in November or December.[16] In an analogous fashion, the farmers in Nazbar share their summer pastures with the Gujur.

To sum up, the people of Yasin have found a wide range of spatio-temporal solutions to combine the cultivation of crops with animal husbandry effectively. That these solutions are subject to long-term changes can be shown by another example. In former times, the upper regions of Qorkultibar were used as high pastures by a number of households from the village of Sultanabad. Their alpine settlement, Ghonomokutunts, was located at an altitude of approximately 3,500 m. It comprised of an agglomeration of small huts for the families and two folds for the animals. In the first half of the 20th century this alpine settlement was

---

14  In the summer settlements, farmers use to let a field lie fallow every two years. In the main villages, by contrast, fallow periods have become rare and are only used by farmers who own more land than the average farmer.

15  This holds true only for families who need not harvest fields in other summer villages. Those who do will go to these sites after the harvest in Ishkaybar.

16  Since the Gujur must feed their animals during the winter mostly with purchased fodder, they stay on the alpine pastures as long as possible.

converted into a summer settlement. On request of the people, the local ruler divided the land between the households and gave permission to cultivate fields. The farmers abandoned now their old huts in favour of separated farms scattered between the fields, each having its own folds and stables. During the 1940s, some farmers, who already owned fields in Ghonomokutunts acquired additional land in Batorkushi, a summer settlement located down the valley some 300 m lower. Due to its milder climate, this site could be used earlier in the year. Thus, the farmers first spend one month in Batorkushi, before they shifted to Ghonomokutunts. However, since the end of the 1960s the fields and pastures in Batorkushi have fallen out of use. Some years after that, especially since the mid-eighties, some farmers have also abandoned their land and grazing grounds in Ghonomokutunts. Until 1991 about a quarter of the farms in this summer settlement had fallen into disuse. Only thirteen families from Sultanabad and three other villages continued to spend the summer in Ghonomokutunts in order to look after their livestock and animals of other households of their villages.

## 5. LEGAL PREREQUISITES FOR THE USE OF HIGH PASTURES

The spatial and temporal co-ordination of crop cultivation and animal husbandry is not only related to ecological criteria; it is also closely linked to various customs and traditional rights, which regulate the utilisation of the high pastures. In this regard, one has to distinguish between the right to graze animals, the right to cultivate crops, the right to stay on the high pastures and the right to cut firewood and timber. A local household may have, for instance, all these rights or only the right to use the grazing grounds and/or to cut wood. Moreover, the right to stay on the high pastures does not necessarily include the right to grow crops. By contrast, the latter two regulations are always linked with the right to graze livestock on the alpine meadows.[17]

What are called rights in this context are not recorded titles but claims legitimated by oral traditions, which are passed on from generation to generation. In this regard, many households can quote a particular legend.[18] However, from a judicial point of view the existing regulations are unsettled. Since the constitution

---

17 The right to stay on the high pastures includes the permission to build a house or hut, while the establishment of fields presumes the right to grow crops. Buildings and fields are considered as private property, which can be sold.

18 By way of example, the *qóom* Shamoné, which mainly lives in Sandi, explains its right to use the high pastures of Asumbar with the following legend: Shah Hamun was an ancestor of the Shamoné. He saved the wife of the local ruler, *tham* Badshah (probably ruling during the 18th century), from enemies during a battle. Shah Hamun took the ruler's wife home in *purdah*, that means without having seen her face. For his brave deed, the local ruler awarded him with gold. Yet, Shah Hamun exchanged this gold against the high pastures of Asumbar. Another popular legend, which circulates on this issue, quotes the extraordinary hunting skills of Shah Hamun. Since the *tham* highly admired his abilities, he presented him Asumbar. By contrast, according to the narrative of a female member of the Shamoné, it was a woman, who received Asumbar as a gift from the *tham*.

of Pakistan does not apply to the Northern Areas and a land settlement has not yet been carried out in Yasin, an official registration of property titles is still pending. Thus, a documentation of the various rights of the local households to make use of the high pastures is overdue.[19]

The pastures, which are located in the vicinity of the villages and which are used in spring, autumn and winter, are generally considered communal land. The grazing grounds of adjacent villages are usually separated, for instance, by minor side-valleys, big stones, huge trees or pyramids of smaller stones, which are made to mark the border. The herds of the different village quarters tend to be taken to distinct pastures within these boundaries. However, this practice does not assign exclusive rights to particular quarters for the use of certain pastures.

The regulations for the utilization of the high pastures are especially complex. Some summer and alpine settlements are used exclusively by the inhabitants of one village. In many cases, however, people of various villages and groups of different descent meet in the settlements on the high pastures. Vice versa, the households of one village are scattered over several summer settlements. These conditions become even more complicated, if it is taken into account that many households possess and use houses and fields in various locations on the high pastures, sometimes even in various valleys. To complicate the situation entirely, many households on the high pastures take care of animals of their relatives and neighbours, who have no rights to stay here during the summer or to cultivate fields in this place.

While it is not possible to relate the individual village to a particular summer settlement, it is feasible to correlate specific sections of the valley with particular high pastures (cf. Fig. 3). In Southern Yasin, the households have only access to few high pastures in some small side-valleys. This is probably one reason why many of these households put their livestock during the summer in the care of relatives who live further north. The inhabitants of the villages between No and Taus use high pastures in the upper parts of Nazbar.[20] The same holds true for the inhabitants of the villages in Nazbar. The central part of Yasin is closely linked to high pastures in Qorkultibar and Asumbar. These eastern side-valleys are, however, also used by villagers of Northern Yasin, who additionally have access to high pastures in Garmashbar. The households of the very north of Yasin have summer settlements in various side-valleys around Darkot. In Thuibar, the high pastures are partly located on slopes near the villages and partly at some distance in side-valleys. This association of different areas of high pastures with distinct parts of Yasin seems to reflect the ancient division of the valley into Salgan (Northern Yasin), Yasin and Thui. Because of the traditional claims involved in

---

19  AZAM ALI (1990) shows for the village of Chaprot in the Nager Valley that this situation may cause conflicts. The governmental forest department as well as the village committee lay claim to the use of the forests, because the former considers the woods as state property, while the latter claims it as communal possession.

20  The Dom, a qóom of low social status to whom the local musicians belong and who live in the villages of Atkash and Bujayot, do not use the high pastures in Nazbar. Due to their status, they have a separate high pasture in Manich-/Bujayotgol.

utilization of high pastures, these divisions still function – in our context – as separate economic regions.

It is difficult to explain the genesis of the complex legal and spatial regulations presented above. In this respect, local informants state two – non-exclusive – theories:

- According to one theory, the *qóom* of the initial settlers of a village (*fatakin*) were entitled to use the adjacent high pastures in every respect. At this time, the members of the descent group and the inhabitants of the village were identical. It is reported, for instance, that the high pastures of Garmashbar were formerly property of the *fatakin* of Barkulti, who are members of a *qóom* with the name Hilbitingé. Over time, this *qóom* split and several members settled elsewhere. Moreover, members of other descent groups established themselves in Barkulti. Thus, the titles related to the utilization of the high pastures were gradually divided among the increasingly heterogeneous population.[21]

- The other theory assumes that land, forests and pastures were the exclusive property of the local ruler. Thus, he decided, which families or entire *qóom* were allowed to use the high pastures and to cultivate fields in the villages and summer settlements. However, the *tham* had the power to withdraw these rights, if the people misbehaved or opposed him. There are several undocumented narratives on the allocation of land and grazing grounds to specific descent groups by the local ruler. The *tham* Shah Burush[22], for instance, is said to have rewarded the *qóom* Begalé from Thui with the land and high pastures of Dasbar (cf. Fig. 3). The example of Ghonomokutunts, where households from Sultanabad received the right to cultivate fields, was quoted above. In both cases, the allocation of land and pastures are considered as gifts (*mehrbani*) from the local ruler, which the *qóom* received for their services and loyalty. The *qóom* Shamoné from Sandi, however, purchased (*sarkharit*) land and pastures in Asumbar from the local ruler.[23] A document from 1906, which is signed by the head of the administration of Jammu and Kashmir in Gilgit, the Wazir Wazarat, proves this transaction. So far, governmental institutions accept this document and hence the property rights of the Shamoné for Asumbar.[24] Yet, some households from Sultanabad contested

---

21   In the course of time, inheritance and household divisions have led to an excessive fragmentation of property titles in Yasin. However, land in the villages and summer settlements usually remain the property of the same – patrilineal – *qóom*, because heritage is passed on only to male descendants. Women are not considered although Islamic law entitles them to inherit property.

22   Shah Burush was the grandson of Shah Khoshwaqt, the founder of the Khoshwaqté dynasty of Yasin rulers originating from Chitral, and lived presumably during the second half of the 18th century. He himself became ancestor of the Burushé, a dynasty ruling the Punial district at the beginning of this century (cf. BIDDULPH 1880: genealogy following p. 154).

23   Cf. footnote 18 n.

24   The document on Asumbar, written in Urdu, reads as follows: "The high pastures of Asumbar from the east to the west on both sides of the Asumbar River [are] inheritance of the *qóom* Shamoné and purchase from the dead *tham* Badshah of Yasin. A golden plough

this document at court. Since wood is less scarce in Asumbar than in other side-valleys, the petitioners tried in this way to obtain access to these high pastures.

After the death of the last *tham* of Yasin in 1967 and the abolition of the local principalities in 1972, the status quo of the customary utilization of the high pastures did not change. The fact that a few households sold some land is the only exception.[25] The legal titles, however, remain uncertain. Thus, there are many disputes between various groups, who claim to have the same rights to certain high pastures. Diverging opinions on the amount of firewood which a *qóom* is allowed to pick up or cut at the high pastures are frequent causes for conflicts.[26] Other reasons for quarrels are violations of national laws, which prohibit the cutting of timber or hunting of big game on the high pastures. Sometimes, such disputes escalate. During the 1970s some unknown persons burned the fields of the *qóom* Begalé in one of the summer settlements in Dasbar thus destroying the entire harvest. The increasing scarcity of land resources in the villages, difficulties in supplying the households with sufficient firewood for the winter, and offences against the honour of a family or *qóom* are the common motives behind such hostilities.

## 6. CURRENT CHANGES IN ANIMAL HUSBANDRY AND THE UTILIZATION OF HIGH PASTURES

Yasin's peripheral location in relation to the urban centre of Gilgit and the economically important Karakoram Highway does not exclude the valley from socio-economic transformation. With the changing socio-economic environment, the conditions change under which the individuals act, which they need to take into consideration or use on their behalf. Among others, these developments

---

and other jewellery [were] given. The pastures of Asumbar from Sulja [located in the Ishkoman Valley; note of the authors] to the Asumbar bridge in Sandi [are] inheritance of the *qóom* Shamoné and [are] its own purchased territory. All claims from other parties are illegitimate. This is written down in the document. The following [persons] are witnesses for Asumbar: Musharafgan Bap [from] Noh Ghizer, Bachdur Bap [from] Noh Ghizer, Molwi Jinjol [from] Teshil Farah, Moghul Bap [from] Tehsil Farah. The document ends here. Wazir Wazarat, Gilgit, Jammu and Kashmir, 15.4.1906, Anwar [signature of the offical]."

25 Some quantitative information on such transactions were obtained by a survey, carried out by Georg Stöber in 50 households in Sultanabad in 1991. Two of these households sold some land in a summer settlement several years ago, while four of them bought land there quite recently. Even today, people often pay in kind. In 1987, for instance, a household gave one cow, one calf and two goats for a field. It has to be kept in mind, however, that the value of land is much lower in the summer settlements than in the villages.

26 It is usually allowed to fetch 12 to 14 donkey loads of firewood from juniper trees per household and year, i.e. approximately 1.4 to 1.7 t. Those households, however, which consider themselves as the owners of particular high pastures, claim an unlimited supply of firewood for their households.

resulted in a number of changes within the system of animal husbandry and the use of the summer settlements. In various regions of the Hindu Kush, Karakoram and Himalaya there has been a general decrease in the number of households using the high pastures. In Yasin, however, we observe both decreasing and increasing trends depending on the aspects we are looking at. Thus, we have to differentiate between cropping and herding activities in the summer settlements. Moreover, we have to distinguish developments on different levels, between changes affecting the practice of the domestic economy on the household level and aggregated figures reflecting changing patterns in Yasin's economic activities in general.

One of the most obvious changes, which we will discuss in detail, concerns the care of animals on the pastures. Taking sheep and goats to pasture is a male duty. In former times, this task was carried out by specialized herdsmen, the *huyeltarts*. Every day, these herdsmen took care of all sheep and goats from the households of their descent group. Only some affluent households with large flocks had their own *huyeltarts*. The herdsmen were usually sons or younger brothers of the head of the family. They received their food every day as well as occasionally extra food (e.g. butter, buttermilk, meat) and other animal products (e.g. skins) from the households whose livestock they tended. The herdsmen had a festival of their own, called *qalamdari*, and specific customs, such as bugling cow horns. Thus, they may be considered as a particular occupational group.

Due to the fact that in recent times a growing number of young men have been attending schools or want to be trained in more "modern" occupations, such as tailoring, the work of herdsmen has grown less attractive. Towards the end of the 1970s, this trade rather suddenly ceased to exist. This process was enhanced by the Aga Khan, the religious leader of the Ismailis, the majority of Yasin's population, who stressed the importance of education in a number of appeals. With the lack of traditional herdsmen, households (often belonging to one neighbourhood) joined together into groups to take turns (*nabat*) herding their livestock. Depending on the number of households involved, each family within a herding group is obliged to take care of the group's goats and sheep every two or three weeks for a whole day. This usually leaves families free to pursue some other occupation or to send their children to school. *Nabat* is applied during the period the livestock is kept in the village, but in some places even during summer. This form of communal co-operation based on a rota system is also found in neighbouring areas, for instance, in Nager and Hunza, where it is known as *naubat* or *galt* (cf. BUZDAR 1988 and KREUTZMANN in this volume). Therefore it seems likely that the *nabat*-system has been taken over from neighbouring areas and suited to local conditions.

The lack of herdsmen could well have been made up for by elderly men. Many of them are strong enough to do the work of a shepherd. This, however, was not taken into consideration, because herding activities are a task of low prestige which seems incompatible with the social status of elderly men.

Other changes regard the practice of farming and herding during the summer months. Up to now, however, trends have not been universal. On the one hand, a decrease can be observed in the utilization of farmland and the time families spend in their summer villages; this development is less marked with respect to the utilization of grazing grounds. In recent times, several households have ceased to send family members to their summer settlements and, in some cases, have abandoned their fields. In the long run, deserted fields and houses are the visible results of these developments. Today, such extreme consequences are still exceptional, although the patterns of utilization have already changed. In some cases, for instance, it is only women and children who stay permanently in the summer settlements. Their husbands or male relatives come only sporadically in order to do some field work or take part in the *nabat*-system where this is applied during summer, too. If the summer settlement is close to the homestead, the fields may be tilled even without taking abode in the summer settlement. This, too, is common in Yasin and keeps some households from being forced to abandon scarce agricultural lands.

In some cases, however, we can observe the opposite trend, i.e. an intensified use of summer settlements, where farmers have cultivated new land. This happened, for example, in Dasbar. A likewise result is also achieved by extending the period of stay. While families in Yasin usually spend four to six months in their summer settlements, at some places they stay far longer. Families from Sandi may dwell up to eleven months in their estates in Asumbar. Visible signs of this change, which particularly takes place at lower altitudes of the high pastures, are the growing numbers of gardens and the construction of large houses with storerooms. These developments may constitute the first stages of the creation of a permanent settlement.

There are several factors at the bottom of these contradictory developments. People stay longer especially in low-lying settlements particularly when they profit from innovations which make access and the life there easier. The construction of roads, for instance, which connect the villages and summer settlements (e.g. Qorkulti village and Tashibar, or Nazbar village and Asqurthan) have improved the accessibility of some summer settlements. In these places, it is now possible to use agricultural machines, such as tractors and threshing machines. These reduce the workload in the summer settlements considerably.[27]

Another factor, which contributes to an intensified use of high pastures results from an increase in the number of livestock. A survey, carried out by Hiltrud Herbers in the villages of Barkulti and Sandi in 1992/93 (59 and 95 households, respectively), proves that the number of animals per household has either remained constant or increased slightly in the course of the past ten years. Thus, the average number of cattle per household rose in Barkulti from 5.4 to 5.7, and that of sheep and goats from 10.0 to even 11.6. In Sandi, the number of cattle increased from 4.3 to 4.7 per household, while the number of sheep and goats has remained constant at 8.3. Likewise, in both villages the number of yaks, donkeys

---

27 However, a negative consequence of the enhanced accessibility is an increased illegal felling of trees. It is not only easier to transport wood by tractors than by donkeys, it is now also possible to obtain much larger quantities than before.

or horses, did not change compared to 1982/83 (cf. Tab. 2). However, since the number of households increased considerably during this period, so did the total number of animals, and some high pastures have already suffered from overgrazing.

But it is not only the high pastures which are affected. As this augmentation goes hand in hand with population growth – the population of Yasin increased from 6,300 in 1911 (Census of India 1911) to approximately 30,000 today – land, fodder and water resources are under stress even more so in the villages. Against this background, people dispose of two strategies: to intensify the use of every agricultural resource possible and to take up off-farm employment. However, very often both strategies conflict as most households do not control a sufficient supply of labour to act in one way and the other. Decisive is the composition of the households. While extended families have usually many adult members, smaller households of nuclear families, whose number is rising, often comprise only of one man, one woman and some children. The lack of labour disables the latter to continue the traditional way of herding and cropping in the villages and summer settlements. Where men have taken up off-farm employment, even larger households may be unable to use all of their agricultural resources. Moreover, a growing number of boys, and more recently, of girls are attending school. Thus, they can neither reduce the workload of their parents considerably nor can they spend the whole summer with an adult relative on the high pastures.

However, due to the expansion of local markets and a better supply of foodstuff, the population of Yasin is nowadays less dependent on subsistence production. People are able to supplement deficits by purchasing food on the market. Furthermore, an off-farm income generally reduces the necessity to cultivate fields that are not easily accessible, such as those in distant summer settlements. The trend to abandon such fields may be strengthened by the fact that at higher altitudes (above 3,000 m) only barley can be grown. As people prefer wheat, which is available on the market the whole year around, the necessity to grow barley has declined markedly.

Additionally, young people are increasingly reluctant to spend their time on the high pastures. In particular, adolescents who went to school are unwilling to pursue the heavy work in the summer settlements (e.g. cutting wood, processing milk). Young people also dislike the monotony of life in the summer settlement, where neither sports nor festivities take place. On the other hand, tensions or conflicts within a household may form an incentive to spend the summer on the high pastures and to stay as long as possible. This holds true especially for many daughters-in-law, since they can lead an independent household in the summer villages without the control of their mothers-in-law. For women, life on the high pastures is, in many respects, less restrictive than in the villages.

In the course of the 20th century, Yasin has lost its political and economic autonomy. The degree of the integration into a larger social and economic system has intensified. People are increasingly dependent on monetary income to meet their needs. In the long run, agriculture seems to be losing its central importance for the reproduction of the household members. In consequence of this development, the practice of combined mountain agriculture is changing significantly. Since an increasing number of men pursue off-farm employment, a shortage of male labour will cause problems. But while most agricultural tasks in the villages can be managed either after the off-farm working-time or by female household members, especially when ploughing and threshing are mechanized, it is more complicated to organize the utilization of the high pastures. However, the recent introduction of the *nabát*-system has proved that the people probably will establish new organizational structures in order to maintain crop cultivation and animal husbandry in various locations. Contrary to the *nabat*-practice, which is a form of mutual co-operation, such solutions may include some kind of professional specialization for certain tasks and a payment in cash for the person who offers his services. In this way, a form of combined mountain agriculture could be maintained, be it as a source of additional income only for many of the households involved. The mode of production, however, will cease to be called a domestic economy.

## 7. REFERENCES

AZAM ALI, Ameneh (1990): Whose Forest is it Anyway? The Harald, Januar 1990, pp. 112–117
BIDDULPH, J. (1880): Tribes of the Hindoo Koosh. Calcutta: Superintendent of Government Printing (Reprint Lahore: Ali Kamran 1986).
BUZDAR, Nek (1988): Property Rights, Social Organization, and Resource Management in Northern Pakistan. Honolulu (Environment and Policy Institute, Working Paper No. 5)
Census of India, 1911, vol. 20, Kashmir, pt.II. Lucknow 1912
CLEMENS, J. and NÜSSER, M. (1994): Mobile Tierhaltung und Naturraumausstattung im Rupal-Tal des Nanga Parbat (Nordwesthimalaya): Almwirtschaft und sozioökonomischer Wandel. In: Petermanns Geographische Mitteilungen 138, 371–387
DURRENBERGER, E. P. (ed., 1984): Chayanov, Peasants, and Economic Anthropology. Orlando: Academic Press (Studies in Anthropology)
GRÖTZBACH, E. (1984): Bagrot – Beharrung und Wandel einer peripheren Talschaft im Karakorum. In: Die Erde 115, 305–321
HASERODT, K. (1989): Chitral (pakistanischer Hindukusch). In: Haserodt, K. (ed.): Hochgebirgsräume Nordpakistans im Hindukusch, Karakorum und Westhimalaya. Berlin: Institut für Geographie der TU, 43–180 (Beiträge und Materialien zur Regionalen Geographie 2)
HERBERS, H. (1995): Ernährungssicherung in Nordpakistan: Der Beitrag der Frauen. Geographische Rundschau 47, 234–239
HERBERS, H. (1998): Arbeit und Ernährung in Yasin. Aspekte des Produktions-Reproduktions-Zusammenhangs in einem Hochgebirgstal Nordpakistans. Stuttgart: Steiner (Erdkundliches Wissen 123)
HERBERS, H., STÖBER, G. (1995): Bergbäuerliche Viehhaltung in Yasin (Northern Areas Pakistan): organisatorische und rechtliche Aspekte der Hochweidenutzung. In: Petermanns Geographische Mitteilungen 139, 87–104
HEWITT, K. (1989): The Altitudinal Organisation of Karakoram Geomorphic Processes and Depositional Environments. In: DERBYSHIRE, E. and L. A. OWEN (eds): Quaternary of the Karakoram and Himalaya. Berlin: Gebr. Borntraeger, pp. 9–32 (Zeitschrift für Geomorphologie N.F. Suppl.-Bd. 76)
HUGHES, R. E. (1984): Yasin Valley: The Analysis of Geomorphology and Building Types. In: MILLER, K. J. (ed.): The International Karakoram Project. Vol. 2, Cambridge, 253–288.
KITSCHELT, F. (1987): Die Kunst zu Überleben. Reproduktionsstrategien von Fischern in Jamai-

ka. Saarbrücken – Fort Lauderdale: Breitenbach (Bielefelder Studien zur Entwicklungssoziologie 34)

KREUTZMANN, H. (1989): Hunza: Ländliche Entwicklung im Karakorum. Berlin: Reimer (Abhandlungen – Anthropogeographie 44)

KREUTZMANN, H. (1995): Communication and Cash Crop Production in the Karakorum: Exchange Relations under Transformations. In: STELLRECHT, I. (ed.): Pak-German Workshop: Problems of Comparative High Mountain Research with Regard to the Karakorum, Tübingen, October 1992. Tübingen: Institute of Social Anthropology, 100–117 (CAK Occasional Papers 2)

LANGENDIJK, M.A.M. (1991): The Utilization and Management of Pasture Resources in Central Ishkoman. Gilgit [unpublished AKRSP study]

MIEHE, S. et al. (1994): Humidity Conditions in the Western Karakorum as Indicated by Climatic Data and Corresponding Distribution Patterns on the Montane and Alpine Vegetation. Erdkunde 50, 190–204

OWEN, L., DERBYSHIRE, E. (1989): The Karakoram Glacial Depositional System. In: DERBYSHIRE, E. & L. A. OWEN (eds.): Quaternary of the Karakoram and Himalaya. Berlin: Gebr. Borntraeger, 33–73 (Zeitschrift für Geomorphologie N.F. Suppl.-Bd. 76)

SAUNDERS, F. (1983): Karakoram Villages. An Agrarian Study of 22 Villages in the Hunza, Ishkoman & Yasin Valleys of Gilgit District. Gilgit, FAO (Integrated Rural Development Project PAK 80/009)

SNOY, P. (1993): Alpwirtschaft im Hindukusch und Karakorum. In: SCHWEINFURTH, U. (ed.): Neue Forschungen im Himalaya. Stuttgart: Steiner, 49–73 (Erdkundliches Wissen 112)

STALEY, J. (1969): Economy and Society in the High Mountains of Northern Pakistan. Modern Asian Studies 3, 225–243

STELLRECHT, I. (1992): Umweltwahrnehmung und vertikale Klassifikation im Hunza-Tal (Karakorum). In: Geographische Rundschau 44, 426–434

STÖBER, G. (2000): Irrigation Practice in Yasin, Northern Areas of Pakistan. In: KREUTZMANN, H. (ed.): Sharing Water – Irrigation and Water Management in the Hindukush – Karakoram – Himalaya. Oxford et al.: Oxford University Press, 72–87

STÖBER, G. (in press): Structural Change and Domestic Economy in Yasin (Northern Areas, Pakistan). In: EHLERS, E. (ed.): Contributions to the Cultural Geography of the Karakorum. Köln: Köppe (Culture Area Karakorum, Scientific Studies 6)

TSCHAJANOW, A. (1923): Die Lehre von der bäuerlichen Wirtschaft. Versuch einer Theorie der Familienwirtschaft im Landbau. Berlin: Parey

WHITEMAN, P. T. S. (1985): Mountain Oases: A Technical Report of Agricultural Studies (1982–1984) in Gilgit District, Northern Areas, Pakistan. Gilgit: FAO/UNDP

# COMING DOWN FROM MOUNTAIN PASTURES
## DECLINE OF HIGH PASTURING AND CHANGING PATTERNS OF PASTORALISM IN PUNIAL

REINHARD FISCHER

## 1. INTRODUCTION

Combined mountain agriculture, comprised of irrigated crop production and animal husbandry, has for long been the established pattern of land use in the Karakorum mountains. The animal manure was used to fertilize the fields and animals were needed to pull the plough whereas the crop residues were fed to the animals during the winter. In summer the animals grazed on high pastures thus making use of different altitudes. Neither agriculture alone nor animal husbandry alone was a viable economic strategy, only the combination of both activities ensured optimal and sustainable use of the environment.

The 1970s saw profound changes in Northern Pakistan. The constuction of the Karakorum Highway made the delivery of subsidized grain from the Pakistani lowlands possible. Self-sufficiency in grain production is no longer the only option for survival. Among the innovations that changed agricultural production chemical fertilizers ended the dependency on animal manure. The effect these changes had were not uniform in all valleys of Northern Pakistan. While Hunza saw a sharp decline in the use of high pastures, Rupal and Ishkoman valleys saw an increased use of high pastures.[1] The aim of this paper is to examine the effect of these changes on animal husbandry in the valley of Punial.

## 2. ECOLOGICAL CONDITIONS FOR AGRICULTURE AND PASTORALISM IN PUNIAL

Punial is situated west of Gilgit. The former principality, ruled from 1800 to 1972 by *rajas* from the Burushe family, is now a tehsil in the Ghizer district. 31 settlements are situated along the Gilgit and Ishkoman rivers in altitudes of 1700 to 2400 metres above sea level (Fig. 1). Rainfall is scarce and all agriculture depends on irrigation.[2] Irrigation water is mostly taken from small tributaries (*nalas*); only two settlements, Golodas and Goharabad take their irrigation water

---

1   KREUTZMANN 1989, CLEMENS & NÜSSER 1994, LANGENDIJK 1991.
2   There is no meteorological station in Punial, but the stations in Gilgit and Gupis, east and west of Punial show 131.7 mm and 133.4 mm of precipitation per year respectively. WHITEMAN 1985: Tab. 1.

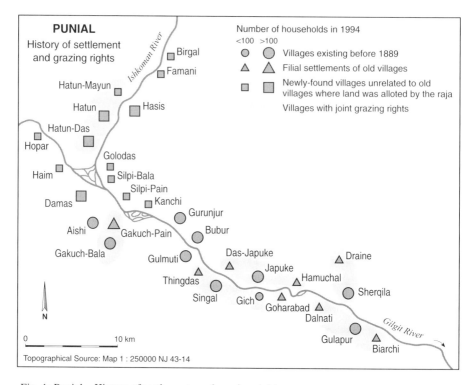

Fig. 1: Punial – History of settlements and grazing rights

directly from the main rivers. Most of the settlements are in double cropping areas with wheat or barley as first and maize as second crop. Only the higher parts of the villages Gakuch-Bala and Aishi above 2300 metres are single cropping areas due to the altitude. The village of Hopar at 2050 metres above sealevel is a single cropping area due to the scarcity of water in spring.

Punial is well-known for its fruits. Grapes, apricots, almonds and walnuts are produced in great quantity.

Ecological condition for traditional agriculture are more favorable in Punial than in single cropping areas. From the end of the 19th century until the mid 1970s Punial produced a surplus of grain that was marketed in Gilgit. From 1889 onward the *raja* of Punial had to supply 250 maunds of grain (= 9.33 t) at a fixed price to the army[3] but he exceeded that amount. In the 1930s he sold 1000 maund (=37.32 t) of grain on the Gilgit market.[4] Throughout the 1950s and 1960s the *raja* continued to sell approximately 1000 maunds of grain on the Gilgit market[5] and in addition to the *raja* some farmers sold their surplus of grain in Gilgit. In contrast to other valleys eg. Hunza, agricultural incomes were sufficient for most

3   IOR/2/1080/255: 69
4   IOL/P&S/12/3285, Gilgit Diary for December 1934

farmers in Punial and the British colonial administrators actually had difficulties to recruit volunteers for the local troops in Punial.[6] Due to migration from other valleys into Punial the population-growth in Punial exceeds the average population growth in the Karakorum-Hindukush region.[7] Throughout this century population-growth in Punial has been tenfold.

Table 1 : Population growth in Punial

| Year | 1880 | 1911 | 1921 | 1931 | 1941 | 1951 | 1961 | 1972 | 1981 | 1995 |
|---|---|---|---|---|---|---|---|---|---|---|
| Population | 2000 | 4423 | 5492 | 6108 | 8164 | 8990 | 11790 | 16826 | 22487 | 35000 |
| Annual Growth | | 2.6 % | 2.2 % | 1.1 % | 2.9 % | 1.0 % | 2.7 % | 3.3 % | 3.3 % | 3.2% |

Source: BIDDULPH 1880, Census of India 1912, 1922, 1932, 1943; Government of Pakistan 1952, 1975, 1984; figures for 1961 cf. Muslim 1990; figures for 1995: author's estimate.

The limiting factor for animal husbandry is the availability of fodder especially in winter and spring. In addition to crop residues from maize and wheat fodder crops like clover and lucerne are grown. These have to be grown on irrigated fields and are thus limiting the available area for wheat and maize.

The slopes above the villages facing North are covered by artemisia steppe which is used as pasture. The slopes facing South are mostly bare of vegetation due to intensive radiation. On the banks of the Gilgit river, areas flooded during the summer are covered with vegetation dominated by tamarisk trees and locally named *mushko*, meaning forest. Especially for the villages around Gakuch on the confluence of the Ishkoman and Gilgit rivers this vegetation is an important pasture during the winter months.

The summer pastures in higher altitudes up to 3500 m.a.s.l. are mostly accessible through paths along the small tributaries (*nala*).

## 3. GRAZING RIGHTS

All land above the highest irrigation channels including the high pastures is considered common property of the village community. The key to membership in the village community is ownership of cultivated lands in the village. Whoever ownes land in the village is entitled to use the high pastures.[8] Some pastures are shared by numerous villages thus reflecting the settlement history of Punial.[9] Filial settlements, that are extensions of older settlements share the grazing right

---

5   Information given by two former tax collectors
6   On the difficulties to find army volunteers in Punial: IOL/P&S/10/826 Gilgit Diary for January 1914 and IOL/P&S/12/3285 Gilgit Diary for March 1935
7   cf. KREUTZMANN 1994: 347 Fig. 8.
8   In Punial like in Ishkoman (LANGENDIJK 1991:25) and Rupal (CLEMENS & NÜSSER 1994: 380) grazing rights are only held by villages, whereas in Hunza (KREUTZMANN 1989:140) and Yasin (HERBERS & STÖBER 1995: 96) grazing rights are held by villages and clans or villages.
9   Cf. FISCHER 1998

with the older settlements. The village of Gulapur and the neighbouring filial-settlement Biarchi for example share grazing rights.

Not all villages possess high pastures. The high pastures in the Gulmuti and Singal *nala* are much bigger than the need of the villages which have grazing rights on them. So the villages controlling Gulmuti and Singal *nala* allow shepherds from other villages in Punial to use these pastures in exchange for a grazing tax.

### 3.1 Grazing tax as part of the tax system

Until 1935 the *raja* of Punial claimed ownership of all lands in Punial including the pastures and collected grazing taxes. In addition to the *gushpur*, members of the *raja*'s family who paid no taxes at all there were three classes of tax-payers.[10] *Bara darkhan* paid 10 maund of grain and 5 ser of *ghi* (butter) annually and had no work obligations to the *raja* exept to accomany the *raja* in war. *Chuno darkhan* payed 5 maund of grain and 2.5 ser of *ghi* annually and had to work on the *raja's* fields and *riyat*, the class of tax-payers on the lower end of the social strata payed a goat per month and had to be ready for all services to the *raja*. In theory the classification of tax-payers did not depend on the size of the landholding or the number of livestock the tax-payer owned, but the classification was reached by agreement between the *raja* and the tax-payer. But it is obvious that those tax-payers with small land holdings unable to produce a grain surplus had to opt for the *riyat* class and thus had to keep large herds to pay a goat per month.

In the context of a tax reform in 1935 the *raja* acknowledged the village communities' right to use the high pastures in exchange for a general tax of 2.5 ser of *ghi* (butter) per household.[11] The amount of grain a household had to pay was fixed in a tax list according to the size of the landholding.[12] In addition each household paid 10 Rupees thus creating the necessity to market part of the surplus.

The *raja* still retained the right to collect grazing taxes from shepherds from outside Punial. Until today the *raja's* clan claims grazing rights in one small part of the Singal *nala*.

After the 1935 tax reform the villages of Singal and Gulmuti controling high pastures used also by shepherds from other villages from Punial got the right to collect grazing tax.

In Gulmuti this tax is still collected directly by the *numberdar*, the elected village headman. In Singal this was done until 1992. From 1993 onward the right to collect the grazing tax is auctioned off to individual members of the village communities. The villagers are thus adopting the same strategy the Pakistani administration uses to organize tax-collection. In theory each shepherd can choose what pasture he uses within the *nala*. But in practice the shepherds from

---

10 IOR/2/1080/255: 169
11 Gilgit Diary for June 1934 in IOL/P&S/12/3285
12 This list *fihrist zamindaran punial* was kindly made available by the *raja*'s family.

Singal occupy the high pastures closer to their village leaving those pastures further away to shepherds from other villages.

### 3.2 Grazing rights for people from Darel and security aspects

The upper parts of Gulmuti and Singal *nalas* bordering Darel in the Diamer District were used by shepherds from Darel who also had to pay a grazing tax. In the first decades of the century shepherds from Darel and Tangir made intensive use of summer pastures not only in Punial, but also in Yasin, Gupis, Kargah and other parts of the Gilgit Agency. Between 1912 and 1916 they did not come to the Gilgit Agency because their ruler Pakhtun Wali did not want them to pay grazing tax. In those years shepherds from the Gilgit Agency produced more *ghi* on the pastures they could now use alone.[13] In 1919 the *raja* of Punial collected 366 goats as grazing tax from shepherds from Darel.[14]

The presence of these people from Darel was perceived as a security threat. The British colonial administration chose not to include Darel and the other valleys of Yaghestan (the land of the rebels) into the Gilgit Agency. Instead these areas were labeled Independent Territories, carefully watched but not interfered by the British. Since 1866 Darel paid an annual tribute in gold to the Maharaja of Kashmir, which was channeled through the *raja* of Punial.[15] The British Political Agent in Gilgit noted all these official contacts between Darel and Punial as well as all crimes like murders and thefts of animals commited by Darelis in Punial.

Recently shepherds from Darel use the upper part of Singal *nala* only on a very limited scale. They are still perceived as a security threat and there is a Police station near the border between Punial and Darel operating during the summer months with officers from Punial.

### 4. CHANGES IN AGRICULTURAL PRACTICES AND THE EFFECT ON ANIMAL HUSBANDRY

The innovations in agriculture available since the 1970s widened the range of economic strategies for the farmers and reduced the interdependence of agriculture and animal husbandry. Chemical fertilizers and mechanization enable the farmers to produce crops without keeping livestock, provided they have the financial means to pay for these inputs. In 1991 32 % of all farmers in Punial used tractors to plough, 76 % used chemical fertilizers and 92 % wheat threshers.[16] In 1995 all wheat harvested in Punial was threshed by machine. In 1993 the last two settlements in Punial, Hatun-Mayun and Draine were linked to the main road by

---

13  Gilgit Diary December 1914 in IOL/P&S/10/826
14  Gilgit Diary September 1919 in IOL/P&S/10/826
15  BIDDULPH 1889: 14
16  PILARDEAUX 1997: 46, Fig. 1

link roads, so that all settlements can now be reached by tractors and threshers. Traditional threshing with animals is very labour intensive in a time of year when the work load reaches its peak. So now all households including those less well-of are willing to pay for threshing by machine.

Animals are still widely used to pull the plough, because most of the fields are very small and not directly connected with link roads. Using a tractor would involve dismantling the walls around the fields and lead to conflicts with neighbours.[17]

Chemical fertilizers came to the Northern Areas as a bulk product in the 1970s. In 1984 the National Fertilizer Cooperation established a fertilizer depot for the Northern Areas in Jaglot on the Karakorum Highway.[18] Today most farmers use a combination of animal manure and chemical fertilizers. Animal manure is perceived soil-improving but much more labour intensive to apply.

## 5. ORGANISATION OF ANIMAL HUSBANDRY

### 5.1 The cycle of seasons

Between the sowing of the first crop in late February or early March and the harvest of the second crop in October animals have to be kept away from the irrigated fields to prevent damage to the crop. During March, April and May the animals are mostly kept in the villages over night and are taken daily to pastures near the villages. From June to September the high pastures up to 3500 m.a.s.l. are used. One station on the way to the high pasture are the summer settlements. These are mostly less than two hours walk away from the villages. Before barley or wheat were produced here and the whole household moved here for a short period of time. Now only less labour intensive fodder crops are grown in the summer settlements and many fields are not cultivated at all. Presently the women and those men who are not performing any duties as shepherds return daily to the village and the houses are abandoned. Only the summer settlement Draine in the Sherqilla *nala* developed into a permanent settlement, inhabited all through the year, whereas all the other summer settlements have lost in importance.

After the harvest of the second crop, mostly maize, in October the animals are led to the fields to graze the stubble residues. From November onward the animals have to be fed with fodder produced on irrigated fields, leaves of willow trees or crop residues. *Nasalo*, the traditional slaughter feast in December reduces the number of animals to be fed during the winter.

Before the 1970s this annual circle was very strict and collectively performed by the village community. Dates were fixed when to go to the high pasture and

---

17 One of the few cases of murder in Punial in the village of Goharabad in 1995 was linked to the fact that two families argued about access to a field for a tractor.
18 PILARDEAUX 1997: 50, PILARDEAUX 1995: 141

when to return to the village. All households had shepherds to accompany the herds to the pastures. The changes and innovations since the 1970s have widened the scope of action for the individual households. Still the village communities fix dates for the beginning of free grazing in the villages after the second harvest and the end of free grazing before the sowing of the new crops. These dates are mostly respected and non-compliances, i.e. accidental grazing of animals on irrigated fields and damage to the crop, is very unusual.[19] But apart from these rules and the ecological restraints there are few restriction. Before there were fixed dates when to use the high pastures. On those high pastures where access rights are undisputed it is now up to the individual shepherd to decide when to go and when to return. Only on those high pastures where access rights are disputed the old fixations and rules are still honoured.[20] Apart from sending a household member as a shepherd with the animals, two other options are feasible today: paying a shepherd from outside the household to take the animals to the high pasture and producing or buying more fodder to feed the animals in the village and to avoid using the high pastures.

### 5.2 Division of labour within the household

The milking of the animals and the processing of the milk to butter in the villages is exclusively done by women. Taking the animals to pastures near the villages is reserved for women or children. Men are accompanying the animals to the high pastures. Among the Gujur minority in Punial women and children join the men on the high pastures. Among the Shina-speaking majority the women do not accompany male shepherds to the high pastures.[21] But women do join the men in the summer settlement in the *nala*.

The fact that many men have off-farm employment and more than 90% of children go to school reduces the number of available shepherds. For those young men who do not have off-farm employment, being a shepherd is not very attractive, because it keeps them away from the village community. Women can not compensate this lack of male shepherds because their main duty is to keep the households in the villages. Only where the workforce in a household is big enough women can be sent for this extra work load.

---

19   In the village of Golodas there can be heard sporadic complaints against Gujurs who only possess little irrigated fields and big herds and who are said not to honour those rules.
20   There is a conflict on grazing rights in Birgal-*nala* between people from Damas who bought land in Birgal-*nala* and claim grazing rights in the *nala* and people from Birgal who want to deny this right to the people from Damas and acknowledge grazing right only for those people who are permanently settled in Birgal. Neither side would use the high pasture before June 1st, the traditional starting date for summer grazing on the high pastures, in order not to be accused of breaking the rules. In Singal *nala* where grazing rights are undisputed there is no official date for the start of summer grazing.
21   cf. SNOY 1993 on the participation of women on the high pastures in the Hindukush-Karakorum area

## 6. ECONOMIC IMPORTANCE OF ANIMAL HUSBANDRY

### 6.1 Sizes of herds

Although it is possible today to practise agriculture without keeping animals all farming households in Punial do still keep animals. Cow milk is an important part of the daily diet. It is consumed as milk-tea and butter is consumed with bread. Milk is a major source of protein since meat consumption is limited to very special occasions in most households. The average number of bovides per household in 1995 was 5.5.[22] This number is higher than in Hunza[23] and comparable with the number of bovides per household in Rupal (Astor).[24] The distribution of bovides is relatively even. There are hardly any households with more than 10 bovides.

The situation with sheep and goats is different. Previously all households kept bigger herds but now there are households who do not keep these animals at all, some households who keep small herds with up to 10 animals, and few households with herds of 100 goats and sheep. In the village of Singal with 260 households only four households have big herds of approximately 100 sheep and goats and only two of the 80 households in the village of Thingdas keep herds of that size.

Asked about the development of herd sizes most farmers answer that the number of cattle increases while the number of sheep and goats decreases, but it is difficult to prove or refute such a statement. No account of herd sizes in Punial was found in the British colonial record. The number of animals kept by a household did not influence the amount of tax the farmers had to pay to the *raja*, so even the *raja* had no reason to keep a record of the number of animals. The earliest account available to the author is an unpuplished agricultural survey conducted in 1978 (Tab. 2).[25]

Table 2: Total number of animals and animals per household in 1978

|        | Total number | Number per household |
|--------|-------------:|---------------------:|
| Cattle | 10312        | 4.3                  |
| Goats  | 12765        | 5.4                  |
| Sheep  | 7019         | 2.9                  |

Source: Tehsildar Office Singal

22  According to the Bovide census conducted by the Animal Husbandry Department Ghizer District in May 1995, 2908 households kept 15,925 bovides. This survey only covered 23 of the 31 settlements with 75% of all households in Punial included. The unpublished results of the survey were made available to the author by the Animal Husbandry Department in Gakuch.
23  KREUTZMANN 1989: 124 Tab. 14
24  CLEMENS & NÜSSER 1994: 384.
25  A file with the results of the survey was kindly made available by the Tehsildar office in Singal.

It can be assumed that these figures are underestimated because most farmers gave smaller numbers of animals in fear of taxation. Taxation on agricultural production was only abolished in 1974 and farmers were not sure about a possible return of taxation either on crop production or on animal husbandry.

SAUNDERS gives the sample figures for total numbers of animals (and animals per household) for three villages in Punial (Tab. 3)

Table 3: Number of animals in three villages in 1983

| Village | Cattle | Goats | Sheep |
|---|---|---|---|
| Gakuch Pain | 570 (6.4) | 535 (6.0) | 312 (3.5) |
| Gakuch Bala | 1256 (7.5) | 1553 (9.3) | 1162 (7.0) |
| Golodas | 351 (5.4) | 695 (10.7) | 438 (6.7) |

Source: SAUNDERS 1983: App. 4

The AKRSP Farm Household Income & Expenditure Survey 1994, conducted on the basis of interviews with ten households per village the number of animals per household in three villages in Punial are given (Tab. 4).

Table 4: Number of animals per household in three villages in 1994

| Village | Cattle/hh | Goats/hh | Sheep/hh |
|---|---|---|---|
| Gulapur | 6 | 9 | 0.2 |
| Aishi | 6 | 11 | 4.5 |
| Japuke | 6 | 17 | 1.1 |

hh = household
Source: AKRSP 1996

It is clear that the number of goats excceds the number of sheep. Goats can make better use of marginal pastures but sheep's wool is better. Due to the availablity of fabricated cotton cloth only few households still process sheepwool to produce cloth.

## 6.2 Marketing livestock

It is a significant fact that although the value of the livestock is a considerable part of all agricultural output only a small proportion is marketed. The AKRSP Farm Household Income & Expenditure Survey 1994 provides a more recent sample (Tab. 5).

Table 5: Share of livestock value and percentage marketed in three villages in 1994

| Village | share of livestock value in agriculture output value | percentage of livestock marketed |
|---|---|---|
| Gulapur | 31.9 % | 7 % |
| Aishi | 37.7 % | 16 % |
| Japuke | 28.2 % | 15 % |

Source: AKRSP 1996

That means that only a small percentage of the livestock is marketed. A limited amount is consumed in the households on special occasions but the biggest share of the herds is seen as a long term investment, especially by those households with small landholdings. Large scale animal husbandry is a risky option because of epidemics that can reduce the herds dramatically within a short period of time. But it is one of the few options for households with small landholdings and limited access to off-farm employment.

Animals can be sold in Gakuch-Pain, the District Headquater. There is a butcher and a considerable demand for meat. But prices are higher in Gilgit, so bigger herds are mostly sold in Gilgit. Transportation to Gilgit is easy to obtain, since most trucks bring goods from Gilgit into Punial and return empty to Gilgit.

## 7. THE GUJUR AS EXPERT SHEPERDS

There are three social groups in Punial. The *gushpur* are the members of the former ruling Burushe family. They paid no taxes before 1972 and still tend to have bigger landholdings than the average households. Some *gushpur* employ sharecroppers to work on their fields and to take care of their herds. The *gushpur* are mostly Sunni Muslim, with one Ismaili branch of the family and speak Shina.

The *zamindar* form the majority of the population. They work on their own fields and paid taxes to the *raja* before 1972. Some *zamindar* work as shepherds with their own herd or with animals of relatives and neighbours. A *zamindar* would not offer his services as a shepherd to strangers. *Zamindar* are Sunni and Ismaili and speak Shina, Khowar and Burushaski.

On the lower end of the social strata are the *gujur*. As a social lable the term does not only include speakers of the Gujur language, but also speakers of the Kilodja language. Some *zamindar* also include the Hindko-speaking households in the village of Famani into this group. All these people migrated to Punial from the south from the end of the 19th century onward, they are all Sunni Muslims and tend to have smaller landholdings than the average *zamindar* household. They are seen as expert shepherds and offer their services by taking herds owned by *zamindar* und *gushpur* to the high pastures. The *gujur* tend to have less formal education[26] and less access to off-farm employment than the *zamindar*. So for

---

26  While almost all Gushpur and Zamindar children receive a primary school education a

many *gujur* being a shepherd is the preferred option especially if they have access to communal pastures.

For *zamindar*, especially for young men beeing a shepherd and using the high pastures is an acceptable but not a very attractive option. Having an off-farm employment and taking part in the social life of the village is far more attractive.

## 8. CONCLUSION

It has been stated that animal husbandry and high pasturing attain greater importance for ensuring subsistence in single cropping areas rather than in double cropping areas.[27] The double cropping area of Punial with its decreasing importance of animal husbandry and high pasturing is a case in point. But animal husbandry and high pasturing are far from disappearing. People in Punial, as throughout the Northern Areas of Pakistan, tend to combine different economic activities in order to minimize risk. High pasturing is still an important economic strategy especially for households with limited landholdings and limited access to off farm incomes.

Comparing the situation in the early 1970s before the construction of the KKH with the situation in the mid 1990s one can see an economic change and a spatial change.

In the mid 1970s the incomes were generated by agriculture in inevitable combination with animal husbandry. A surplus of grain was produced. Off-farm employment was of limited importance. In the mid 1990s off-farm employment in Punial or outside is crucial. Agriculture produces only half of the grain consumed and animal husbandry is practiced on a small scale by most household but on a larger scale with high pasturing only by a few households.

The spatial change is visible in the summer settlements. In the early 1970s households moved for the summer month to the summer settlements and produced wheat, barley or maize there. Now there are only less labour intensive fodder crops produced on those fields and no household moves there any more. The shepherds pass through the summer settlements on their way to the high pastures. The only exeption is the former summer settlement of Draine in Sherqilla *nala*, which developed into a permanent settlement in the 1960s.

In the mid 1990s there is a concentration of all activities in the village. New houses, shops and schools have been built limiting the area available for agriculture. All irrigated land in the villages is used intensively. Most villages have schools, shops and small hotels now.

For young men it is not very attractive to leave the village for a few month to take the animals to the high pastures. For them it seems more attractive to leave the village for Gilgit or the Pakistani lowland to complete their education or find an off-farm employment in order to return to the village to build a new house.

significant number of Gujur children are not attending school, reducing their future chances to find an off-farm employment.

27  CLEMENS & NÜSSER 1997: 242

## 7. REFERENCES

a) Original Document kindly made available by members of the Burushe family in Punial
fihrist zamindaran Punial (list of landowners in Punial): tax-list compiled in 1935.

b) Archival Material from the India Office Library & Records

IOR L/P&S/10/973: Gilgit Agency Diaries 1921–1930.
IOR L/P&S/12/3285: Gilgit Agency Diaries 1912–1920.
IOR R/2/1080/255: Complicity of Khan Bahadur Muhammad Akbar Khan of Punial in two murder cases. 1904.

c) Secondary sources

AKRSP (1996): Farm Household Income and Expenditure Survey 1994: Socio-Economic Profile of the 22 Villages Surveyed in Gilgit. Gilgit.
BIDDULPH, J. (1880): Tribes of the Hindookoosh. Calcutta: The Superintendent of Government Printing [Reprint: Karachi: Indus Publications. 1977].
CENSUS OF INDIA (1912): Census of India, 1911. Vol. 20: Kashmir. By Matin-uz-Zaman Khan. Lucknow: Naval Kishore Press.
CENSUS OF INDIA (1923): Census of India, 1921. Vol. 22: Kashmir. By Chaudhari Khusi Mohammad. Lahore: Mufid-i-Am Press.
CENSUS OF INDIA (1933): Census of India, 1931. Vol. 24: Jammu and Kashmir State. By Rai Bahadur Anand Ram & Hira Nand Raina. Jammu: Ranbir Government Press.
CENSUS OF INDIA (1943): Census of India, 1941. Vol. 22: Jammu and Kashmir. By R.G. Wreford. Jammu: Ranbir Government Press.
CLEMENS, J. & M. NUSSER (1994): Mobile Tierhaltung und Naturraumausstattung im Rupal-Tal des Nanga Parbat (Nordwesthimalaya): Almwirtschaft und sozioökonomischer Wandel. In: Petermanns Geographische Mitteilungen, 138, 371–387. Gotha.
CLEMENS, J. & M. NUSSER (1997): Resource Management in Rupal Valley, Northern Pakistan. The Utilization of Forests and Pastures in the Nanga Parbat Area. In: I. STELLRECHT & M. WINIGER. Perspectives on History and Change in the Karakorum, Hindukush, and Himalaya. Rüdiger Köppe Verlag. Köln. (= Culture Area Karakorum, Scientific Studies 3): 235–263.
EHLERS, E. (1995): Die Organisation von Raum und Zeit- Bevölkerungswachstum, Ressourcenmanagement und angepaßte Landnutzung im Bagrot/Karakorum. In: Petermanns Geographische Mitteilungen, 139: 105–120. Gotha.
FISCHER, R. (1998): The History of Settlement in Punial, Northern Areas of Pakistan, in the Nineteenth and Twentieth Century. In: I. STELLRECHT (ed.). Karakorum – Hindukush – Himalaya: Dynamics of Change. Rüdiger Köppe Verlag. Köln. (= Culture Area Karakorum, Scientific Studies 4/1): 511–526.
GOVERNMENT OF PAKISTAN (1952): Census of Azad Kashmir, 1951: Azad Kashmir, Gilgit & Baltistan (Compiled by Iftikhar Ahmad). Report & Tables. Murree.
GOVERNMENT OF PAKISTAN (1975): District Census of Pakistan: Gilgit District 1972. Islamabad: Census Organization.
GOVERNMENT OF PAKISTAN (1984): 1981 District Census Report of Gilgit. Islamabad: Population Census Organization.
HERBERS, H. & G. STÖBER (1995): Bergbäuerliche Viehhaltung in Yasin (Northern Areas, Pakistan): Organisatorische und rechtliche Aspekte der Hochweidenutzung. In: Petermanns Geographische Mitteilungen 139, 87–104. Gotha.
KREUTZMANN, H. (1989): Hunza. Ländliche Entwicklung im Karakorum. Dietrich Reimer Verlag. Berlin (= Abhandlungen-Anthropogeographie des Geographischen Instituts der Freien Universität Berlin, 44).

KREUTZMANN, H. (1994): Habitat Conditions and Settlement Processes in the Hindukush - Karakorum. In: Petermanns Geographische Mitteilungen 138, 337–356. Gotha.

LANGENDIJK, M.A.M. (1991): The Utilisation and Management of Pasture Resource in Central Ishkoman. Gilgit: AKRSP.

MUSLIM, M. (1990): The Way of Life in Ponyal Valley. Unpublished M.A.Thesis. Peshawar. Pakistan Study Centre Peshawar University.

PILARDEAUX, B. (1995): Innovation und Entwicklung in Nordpakistan: Über die Rolle von exogenen Agrarinnovationen im Entwicklungsprozeß einer peripheren Hochgebirgsregion. Verlag für Entwicklungspolitik Breitenbach. Saarbrücken. (= Freiburger Studien zur Geographischen Entwicklungsforschung, 7).

PILARDEAUX, B. (1997): Agrarian Transformation in Northern Pakistan and the Political Economy of Highland-Lowland Interaction. In: I. STELLRECHT & M. WINIGER. Perspectives on History and Change in the Karakorum, Hindukush, and Himalaya. Rüdiger Köppe Verlag. Köln. (= Culture Area Karakorum, Scientific Studies 3): 43–57.

SAUNDERS, F. (1983): Karakorum Villages. An Agrarian Study of 22 Villages in the Hunza, Ishkoman & Yasin Valleys of Gilgit District. Gilgit: AKRSP (= A technical Publication of the FAO Integrated Project for Rural Development PAK/80/009).

SNOY, P. (1993): Alpwirtschaft im Himalaya und Karakorum. In: SCHWEINFURTH, U.(Ed.), Neue Forschungen im Himalaya. Stuttgart: Franz Steiner (= Erdkundliches Wissen, 112), 49–73.

WHITEMAN, P.T.S. (1985): Mountain Oases: A Technical Report of Agricultural Studies in the Hunza, Ishkoman & Yasin Valleys of Gilgit District. Gilgit: Department of Agriculture/ Integrated Rural Development (= FAO/UNDP PAK 80/009).

# PASTORALISM IN THE BAGROT:
# SPATIAL ORGANIZATION AND ECONOMIC DIVERSITY

ECKART EHLERS

## 1. INTRODUCTION:
## RURAL HOUSEHOLDS AND ECONOMIC DIVERSITY

Economic diversity is the traditional basis of subsistence in rural households of many developing countries. One single activity – be it agriculture, animal husbandry or forestry – is seldom enough sufficient to ensure the minimum income requirements of a rural family. Therefore, combinations of different economic activities are common and usual. Such an observation holds even more true for harsh and fragile environments. Diversity is essential in order to minimize ecological vulnerability, the impacts of natural hazards and of economic risks.

The mountainous Northern Areas of Pakistan are such a harsh environment. Traditional human responses to nature's challenges are a vivid testimony of human creative adjustments to such challenges: it is not only the diversity of economic activities, but also the sometimes admirable strategies to cope with floods and water scarcity, with steep slopes and poor soils, with avalanches and rockfall, with mudflows and erosion. The introductory essay to this volume reflects international research on potentials and constraints of those mixed economies of montane societies. This volume itself focuses on various adaptations of pastoralism of high mountain economies in northern Pakistan. The following remarks, however, refer exclusively to the organizational flexibility of pastoral strategies within the larger context of combined mountain agriculture of rural households in one specific valley, the Bagrot, and as part of the overall rural subsistent household economy.

For the purpose of this paper it is suggested to define a traditional subsistent rural household in the Bagrot valley as a system. Each household or system consists of at least three subsystems which are: agriculture, animal husbandry and – to a minor extent – forestry. None of these subsystems, however, has been self-reliant and sustainable in itself; on the contrary: all of them were closely interlinked and interconnected. It is the hypothesis of this paper that

> for the vast majority of traditional rural households in the past neither agriculture nor pastoralism nor forestry alone were sufficient to sustain a traditional rural household; only the combination of at least two of the three components ensured subsistence.

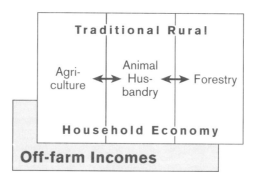

Fig. 1: Traditional rural household incomes in Bagrot

Reality and history show that in the past isolated valleys like Bagrot actually were a kind of economic and social microcosms of their own. Connected to the "outer world" by only a few and extremely vulnerable footpaths as late as the early 1970ies, Bagrot's earlier development is almost exclusively based on the carefully organized and coordinated use of is limited agricultural and silvicultural resources. The fact that Bagrot's society and economy has been a predominantly subsistent one and that outside influences as well as non-rural sources of income were practically non-existent is convincingly shown by SNOY's extensive documentation of this valley in the mid-1950ies (SNOY 1975). Fig. 1 therefore depicts not only the traditional situation of Bagrot's rural households in the recent past, but may also serve as a theoretical framework in comparison to today's situation (cf. Fig. 7).

Both agriculture and animal husbandry developed a set of activities which, in a system-oriented interpretation, may be considered as variables of the corresponding subsystems. In agriculture, grain and fodder production, horticulture or fruit cultivation (apricots!) may be considered as such variables. "Forestry" in the past served exclusively the households' needs as building material and firewood. Animal husbandry and its variability, flexibility and diversity shall be the focus of the following lines.

## 2. PASTORALISM AND ANIMAL HUSBANDRY AND THEIR INTEGRATION INTO RURAL HOUSEHOLD ECONOMIES

In an earlier paper (EHLERS 1995), it was argued that the organization of time and space in the Bagrot valley shows rural households' marvelous command of a rational adaption to the necessities of a harsh environment. Agriculture and animal husbandry are closely interconnected and form a symbiosis that enabled the overall system "household" to generate a sustainable income. The combination of the two subsystems "agriculture" and "pastoralism", however, conceals the indicated fact that each of them is split up into a range of strategies or variables themselves. With other words: in line with the basic hypothesis of this paper, not only agriculture, but also animal husbandry is divided into different strategies to cope with the insecurities of nature and environment.

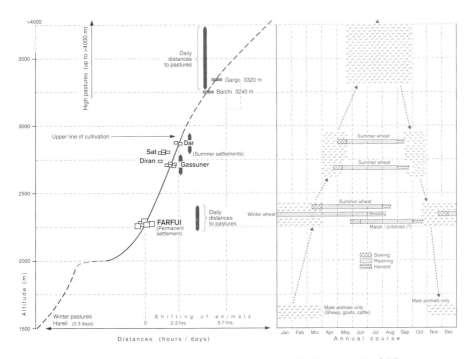

Fig. 2: The use of space and time in agriculture and animal husbandry in Farfui/Bagrot

In regard to animal husbandry and pastoralism as its economic basis it will be shown that its spatial organization as well as its economic diversity are far from being simple. Besides, their integration into the overall rural household activities are remarkable. The case of Farfui/Bagrot (Fig. 2) may serve as an example not only for the close and intimate interactions between agriculture and animal husbandry, but also for the indicated flexibility of the Bagrotis' pastoral management capacities. Animal husbandry is performed on basically four different altitudinal (and regional) levels:
- on the permanently used valley bottom with its villages and open fields
- on the periodically used areas of the summer villages with their arable land and pastures
- on the periodically used high pastures and
- on additional winter pastures outside the Bagrot valley, at least for parts of the herds.

Those generalizing observations, however, vary from household to household and are dependent on a set of variables which include among others:

- household or family size as a determinant for the availability of pastoral labour-force;
- acreage/hectarage of own land as a precondition for additional pasturing during intermediate seasons (esp. in spring and fall) and for the production of winter fodder;
- availability of and access to "communal" pastures in the high regions;
- size and composition of the flocks (sheep, goat and/or cattle), which has direct impacts on the
- productivity and production goals of the subsystem "animal-husbandry".

Especially the family size and the availability of an appropriate labour force are decisive factors not only in regard to the overall economy of the household, but also to its ultimate production goal. HERBERS (1998) and STÖBER (1998) have shown impressively the role and importance of animal husbandry for both the subsistence of the households themselves, on the one hand, and for market production and cash income, on the other. A more general representation of crop farming and animal husbandry, their interactions and their contributions to subsistence or market is contained in Kreutzmann's Fig. 2 in the introductory article to this volume.

The following discussion will concentrate on two aspects of the subsystem "animal husbandry" within the overall rural production system of the Bagrot valley:
1. its spatial organization
2. its economic diversity.

It is based on observations in the years 1993 and 1994, complemented by a few updates during a short visit in the summer of 1999. The focus on the high pastures of Gargo and Sinakar is justified in view of the fact that these examples are more or less representative for the overall situation of Bagrot's pastoralism and which – at the same time – reflect potentials (Gargo) and limitations (Sinakar) of the pastoral subsystem.

## 3. PASTORALISM: ITS SPATIAL ORGANIZATION

One of the main characteristics of Bagrot's animal husbandry is its variability, flexibility and diversity. This general observation holds true not only in regard to the size and composition of herds, but also in regard to the production goals of the animal owners. A decisive factor for such differentiations is the availability of and the access to high pastures in the Bagrot valley. The villages of Farfui/Chirah and Bulchi, located in the upper reaches of the Bagrot, own not only several summer villages, but also – and even more so – dominate comparatively large tracts of undisputed and unchallenged mountain pastures beyond and above their seasonal villages at the valley end. Datuchi and Sinakar, on the other hand, are at a great disadvantage due to their location in the middle and lower sections of the valley. It is not only the restricted or lacking access to irrigation water, but also the comparatively limited land areas that hamper both their agriculture and their animal husbandry (Fig. 3).

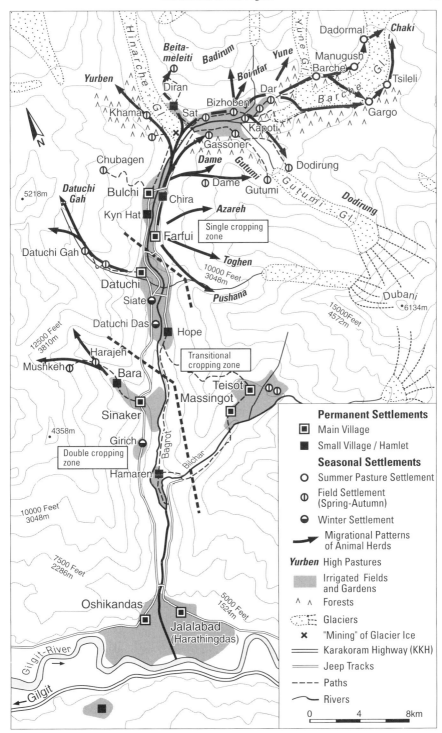

Fig. 3: The spatial and temporal organization of agriculture and animal husbandry in the Bagrot Valley

Gargo and the other high pastures in the northeastern sections of the Bagrot are the common grazing grounds of Farfui/Chirah and Hope. These Bagrot villages claim the high pastures of Pushana with approximately 10 stables, Toghen (2), Azareh (6), Dame (3), Gutumi (5), Gargo (23), Tsileli (10) and Chaki as their exclusive grazing grounds, while Dadormal (3 stables), Manugush (4), Barche (3), Boinfar, Badirum and Hinarche are shared with Bulchi (Fig. 3). While some of these shares seem to be not unquestioned and disputed, especially in regard to the claimed 2:1 share between Farfui/Chirah/Hope, on the one hand, and Bulchi, on the other, it is clear that also Bulchi has ample areas for pastoral use and for summer villages at its disposal. These are – besides those shared with Farfui – Yurben (3 stables), Nali (2), Tagharbari (2), Qoridja (3) a number of smaller pastures such as Gesh, Darabehi, Surghen and others with one or two stables each. Only of Chubagen's altogether 6 stables one belongs to Datuchi.

In contrast, scarcity is the prevailing characteristic of the villages in the central part of the Bagrot. Shortage of land and/or water are common, property rights nevertheless clear and seemingly undisputed. This holds true also for the summer pastures of both Datuchi and Sinakar. Both villages have only small grazing areas above their home settlements. Summer pastures and number of stables are limited: Sinakar has about 16–18 stables (*harajeh*) altogether which, however, are smaller than those of the aforementioned upper reaches of the Bagrot valley. In view of the limited grazing areas above Sinakar, stables do not form clusters, but mostly are isolated and differentiated as lower, middle and upper stables with locations of about 2800, 3300 and 3800 metres, respectively, above mean sea-level. Basically the same holds true for Datuchi with a total of 14 to 18 stables, most of them dispersed over a narrow stretch of slopes above the village and covering heights between 2500 and 4000 m. Both villages claim that expansion and intensification of animal husbandry has always been hampered by the limited availability of appropriate land for pasture.

Unlike the permanent villages themselves and unlike the seasonally used summer villages stables are not very conspicuous structures. Fig. 4 gives a typical example. Mostly built as a unit of stable plus shelter for the shepherds, very often harajeh are carved out of the debris and the gravels of slopes and moraines, thus hiding the bulk of the structure subsurface. The combination of human shelter, stable and corral are typical for almost all these structures. Only in the high pastures of Gargo or Tsileli, stables and human shelters form clusters along the slopes of the mighty Barche glacier; Gargo even has a small mosque.

Availability or non-availability of appropriate pastures determine the spatial organization of their uses. Villages with ample pastures such as Farfui or Bulchi offer choices and options: daily grazing is stretching not only vertically, but also horizontally over wider areas. In recent years many stables even remained empty. Consequently, pastoral productivity is comparatively high. The opposite holds true for disadvantaged communities such as Datuchi or Sinakar: limited space not only determines the smaller sizes of herds, but has also led to overgrazing and consequently to environmental degradation of the pasture lands.

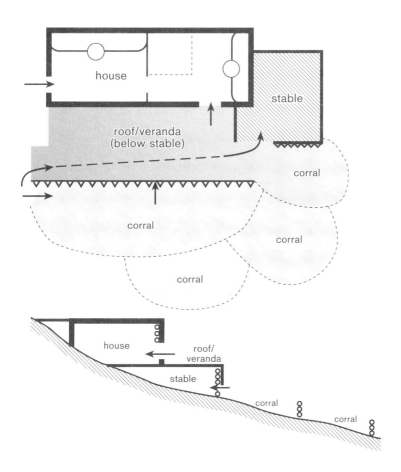

Fig. 4: Typical structure of a "harajeh" (middle stables of Sinakar)

Advantages or disadvantages are increased by the unequal availability of those complementary components that are essential for the animal husbandry subsystem: the transitional pastures and grazing grounds located mostly in or close to the periodically used summer villages and the winter stables. Again, Bulchi or Farfui (Fig. 2) are generously equipped, while Datuchi and Sinakar have merely limited options also in these respects. Only in regard to the scarcity of winter pastures in and around the home settlements all Bagrot villages seem similar.

What are the responses of the Bagroti in regard so these challenges? In a simplified and focused way, one could argue that – *ceteris paribus* – the Bagrot developed admirable strategies to cope with the problems of scarcity creating a wide range of vertical and horizontal pastoral migration patterns and by combining verticality and horizontality into an extremely flexible set of combinations.

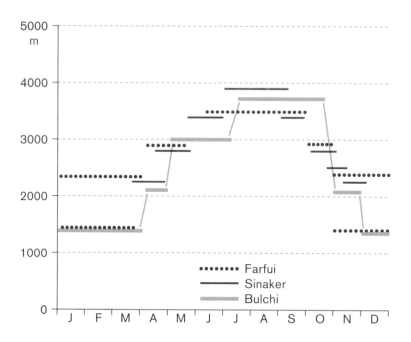

Fig. 5: Pastoralism in the Bagrot. The vertical dimension

Basic patterns of <u>verticality</u> in Bagrot's pastoralism are reflected in Fig. 5. Its main features are ecologically enforced differentiations of Farfui's and Bulchi's pastoralism, on the one hand, and that of Sinaker, on the other. The comparatively good equipment of Farfui and Bulchi with high pastures as well as with summer villages and their grazing grounds enables the herd-owners to spend longer periods in the high pastures resp. in the summer villages, while animal husbandry in Sinaker (as well as in Datuchi) has to adjust to nature's limitations by smaller herds and shorter grazing periods in the various relays of the system.

Variability, flexibility and diversity of the pastoral subsystem become, however, even more apparent in regard to the <u>horizontal</u> pattern of Bagrot's animal husbandry. It is not only reflected in the traditional inclusion of pastures outside the Bagrot valley into the pastoral grazing cycle, but also – and even more so – by splitting up the summerly composition of the herds. Fig. 6 shows both strategies: the traditional uses of grazing rights in Oshikandas (cf. Fig. 3) and especially the inclusion of the Hareli pastures about 6 km west of Gilgit. Both options are important preconditions for the maintenance of a winterly minimum size of the herds. Equally important, however, is the division of the herds. Cows as well as lactating goats and sheep are kept close to the home villages. Here, they graze fallow land and natural pastures around the settlements. Under harsh weather conditions they are kept in stables. On the other hand, several thousand animals are driven in December to Hareli, where they hibernate until end of April. Water

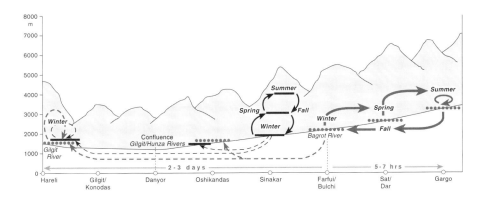

Fig. 6: Pastoralism in the Bagrot: The horizontal dimension

shortage in summer makes Hareli useless as grazing area during the hot season; snow and episodic rainfall provide sufficient vegetation and water for the animals' survival between December and April.

The diffusion of verticality and horizontality into a variable set of combinations is surely the predominant feature of the spatial organization of Bagrot's pastoralism. Fully embedded into the agricultural year, the pastoral subsystems is definitely limited by the availability resp. non-availability of grazing land. Especially winter fodder is crucial. The fact that the hibernation of the animals is the main problem of all forms of animal husbandry in the Bagrot holds true today as 45 years ago (SNOY 1975, 103). Winter forage as limiting factor may also be one of the main reasons that nowadays quite a few of the "harajeh" – especially those of Farfui and Bulchi – remain empty during the summer months. Seasonality and territoriality show a wide range of options – not only interpreted, but realistically seen by the local farmers and shepherds as possibilities and necessities of creative adjustments to the ecological environment and socio-economic framework conditions. These forms of flexibility – reaching from non-exploitation of available resources to overgrazing and environmental exhaustion – change in space and time. Presently, the availability of non-agricultural and non-pastoral alternatives tend to lessen the stress on the pastures of the Bagrot.

## 4. PASTORALISM: ITS ECONOMIC DIVERSITY

Cattle, sheep and goats form the pastoral trilogy of the Bagrot valley. Almost every rural household has varying numbers and differing combinations of their composition. All animals have their specific role and function not only within the animal husbandry subsystem, but also within the overall system of the rural household. Thus, cattle do not only serve as draft animals for ploughing or threshing, but also provide valuable manure for the fields. Milk, meat, skins are important components of subsistence and marketing. Sheep provide wool and are one of the major producers of cash income due to their high fertility and due to the fact that they are breeding twice a year. Goats, finally are not only suppliers of milk as the basis of a number of derivates, but they are highly esteemed due to their almost total unpretentiousness in regard to pastures and fodder supply. Since these very general characteristics are hardly suitable to support the basic view that also the economic diversity of animal husbandry is part of the overall strategy of Bagrot's rural households to cope with ecological vulnerability and with economic risks, it may be appropriate to give further details about pros and cons of the trilogy's components.

Cattle, although constituting the smallest part of the trilogy in terms of number, offer the greatest variety of economic services to their owners. Bulls and oxen are kept predominately as working animals. Less in number than cows, they altogether provide valuable manure as well as meat and skins if they are slaughtered. More important, however, is their role as providers of milk. Outside their breeding and lactating period cows provide 2–4 kg milk per day which is used either for the own subsistent demands of the rural household or - in the case of high pasture production – as a basis for a number of milk derivatives (cf. figures in the articles of EHLERS-KREUTZMANN or KREUTZMANN in this volume!). In these cases cow milk is normally mixed with those of goats. Positive aspects of cattle husbandry are their overall broad production goals and productivity spectrum: they reach from manure to high quality milk derivatives. Milk e.g. is the basis of a number of products, reaching from cheese and butter to qurut and buttermilk. On the average, 12 to 14 kg of milk produce 1 kg of butter of 2 kg of cheese. The price of butter, which in 1993 was 160 Rupees/kg, has risen to app. 240 to 250 Rs./kg in 1999. – Of course, cattle husbandry also has a number of great disadvantages such as:
– great demand for human labour force to guard the animals, esp. in and around the home villages;
– necessity to provide forage and winter fodder for those animals that are stabled during winter;
– higher and more selective fodder demand than goats and sheep.

Besides, introduction of mechanic plows and tractors (for both ploughing and threshing!) as well as chemical fertilizers (cf. PILARDEAUX 1995, 1997) tend to create serious competitions and cheaper alternatives, esp. for the keeping of bulls and oxen.

Goats, dominating in terms of number, are far spread not only among the rural households of mixed mountain agriculture, but also in urban centers such as Gilgit. Undemanding in their fodder supply they nevertheless are good producers of fat quality milk. Even after birth of their lambs, she-goats are prevented of feeding them in order to secure their high quality milk for the production of butter. Once a year, mostly in May/June, goats are sheared; goat-hair being the basis of a number of local products such as rugs, bags, cords etc.. Finally, also goat meat is a major source of additional income, esp. in connection with religious festivities or family ceremonies such as weddings. Skins are sold.

The advantage of goats include their easy maintenance, their versatile and comparatively high yields of fat milk which – beyond domestic use – is almost exclusively used for butter production. However, their seemingly advantage is disadvantage at the same time: the ecological impacts on the environment are considerable. Especially goats' tendency to feed on young sprouts and bark of trees causes considerable damage. Around villages and fields these damages are counterbalanced by simple, yet expensive protection of young trees by villagers.

Sheep, finally, are the least important species in regard to milk production. It basically does not exist. Reason for this is that female sheep give birth to two or three lambs/year, making them a major producer of live animals and meats. As a rule one may say that older sheep, esp. rams, are being sold parallel to the number of animal births of lambs so that the overall number of sheep remains stable. Thus, sale of sheep constitutes a major source of income to Bagrot's rural households. Wool is the second major product of sheep holding: sheared twice a year (in May and September), one kg of wool generates 50–60 Rp.; top quality may even go up as high as 75 Rp./kg (1999). Due to the fact that sheep – much more than goats – tend to be kept in winterly stables they also serve as producers of valuable manure which in spring is distributed over the fields.

In terms of advantages/disadvantage sheep are more demanding than goats in regard to their pastures; ecologically, however, they are less aggressive. While live animals and meat produce cash incomes, their wool is basis for a number of local and domestic products, some of which even of touristic interest. A certain strain is caused by the winterly keeping animals in stables, esp. of female and young sheep. They demand not only an appropriate provision of winter forage, but also supervision by villagers in times of free grazing in and around the stables.

Altogether, cattle, goats and sheep are not only basis of a wide range of additional incomes for the overall rural households system of the Bagrotis, but the subsystem pastoralism /animal husbandry creates many and meaningful overlaps with the subsystem agriculture. These overlaps become esp. apparent in connection with the use of agricultural lands:
- in the summer villages, fields before sowing and after harvest serve as grazing areas for animals (esp. cows) on their way to and from the high pastures. The use as pastures – three to four weeks in spring and autumn – means at the same time a natural fertilization of these fields;
- in the home villages the availability of the harvested fields for 4-5 months of additional grazing land (*hedd*) has the same benefits to both farmers and herdsman.

These symbioses between animal husbandry and agriculture, however, are not the very essence of their practice. Again: neither agriculture nor animal husbandry alone are able to maintain and sustain a traditional rural household. Economically, agriculture and animal husbandry belong together, supplemented by forestry as an additional means of subsistent income. In view of these facts, it is legitimate to interpret not only spatial variability, but also and likewise economic diversity of animal husbandry as a response strategy to ecological vulnerability and economic risks. As a matter of fact: again and again, Bagrot farmers stress the necessity of diversity in order to cope with the harsh environment and its versatile natural hazards, not to speak of the increasing impacts of market conditions and other external factors on their decisions.

## 5. THE FUTURE OF BAGROT'S PASTORALISM: FROM SUBSISTENCE TO MARKET

This is neither the place nor the opportunity to discuss Bagrot's history (cf. SCHNEID 1997, SNOY 1975) nor the details of the valley's recent transformations. For the sake of this topic it must suffice to state that neither Pakistan's independence and statehood in 1947 nor the abandonment of northern Pakistan's independent kingdoms and fiefdoms in 1974 have been the decisive factors for the transformation of Bagrot's rural economy. It was the completion of the Karakorum Highway (KKH) in 1978 and the construction of jeep-tracks into the Bagrot valley. In 1971, first jeeps could go as far as Sinakar, 1976 both Bulchi and Farfui could be reached. Today, all villages in the main valley are connected with Gilgit, the capital of Pakistan's Northern Areas.

The opening of Bagrot to the outside world and the permanent and regular connections of its villages to Gilgit and beyond have changed attitudes and behaviors of the villagers as well as the traditional economy of the valley. The establishment of schools (SCHNEID 1997), the access to daily newspapers and the wide spread of transistor radios, especially, however, the installation of electricity in 1994 and, as a consequence, the arrival of first TV-sets have caused – and are still causing – cultural disruptions. Education, job opportunities and different life styles have instigated esp. young males to leave the valley and to look for better opportunities in Gilgit or in Pakistan proper.

Equally important, however, are economic transformations. The accessibility of Bagrot for tractors and threshing machines have changed traditional farming techniques, new varieties of seeds and the introduction of chemical fertilizers the traditional agricultural production (PILARDEAUX 1995, 1997). Equally dramatical are changing attitudes towards food production and consumption. DITTRICH (1995, 1997) has shown in detail the impacts of dietary changes as a result of food imports – and the population's vulnerability to these changes as a consequence of natural hazards along the KKH and the interruptions of food supplies to the Northern Areas.

Finally, the opening of Bagrot to the outside world has resulted in a great variety of different off-farm incomes. As shown elsewhere in more detail (EHLERS 1995, SCHNEID 1997), recent economic activities of traditional rural households include – among others – incomes from labour for governmental and non-governmental organizations, remittances from different forms of off-farm labour in the region, in other parts of Pakistan or abroad, or revenues derived from local trade and commerce. In quite a few cases, these new sources of income are not only much higher than those derived from traditional agricultural and/or other

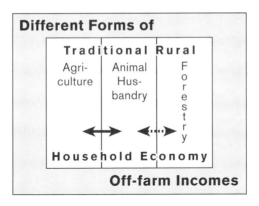

Fig. 7: Modern structure of rural household incomes in Bagrot

rural activities, but sometimes they are so substantial that all traditional forms of a mixed mountain economy are not competitive any more and tend to become of minor importance or even irrelevant. This holds especially true for the role and function of traditional forestry: while it is still of importance to most rural households, the unwarranted commercial exploitation of Bagrot's forests is dominated by two or three families, thus, loosening the ecological and economical ties between the traditional subsystems of Bagrot's rural households.

And what about Bagrot's future agriculture and animal husbandry? There is no doubt that – contrary to the recent past – Bagrot's agriculture and animal-husbandry are today part of a larger economy remittances from Bagroti working in the Northern Areas, in Pakistan or in the Gulf states, pensions for former soldiers and other government employees, income from trade and seasonal work as well as an increasing number of government employees and other servicemen in the valley itself contribute to a wide spectrum of job and income opportunities.

Improved accessibility of Bagrot and resulting changes in the socio-economic framework conditions of its population have contributed to the fact that both agriculture and animal-husbandry have attained a new role and importance in the overall economy of the valley (Fig. 7). Both are still major providers of a wide range of basic nutritional requirements of the majority of Bagrot's rural households. On the other hand, it is beyond any doubt that the innovative use of agricultural machinery and the increasing exodus of young, mainly male Bagrotis contributes to major shifts in the temporal and spatial organization of both agriculture and pastoralism. Already now, the lack of manpower – especially during the busy summer season – has led to abandonment of a few fields and to the reduced sustainable exploitation of the high pastures, especially those of Bulchi and Farfui.

Selective abandonment of marginal lands and pastures, however, is both an ecological and economic chance for harsh and fragile environment (KUHNEN 1992, PRICE & THOMPSON 1997). In line with Fig. 4 of the introductory article to

this volume, it is therefore suggested to develop a new equilibrium which takes into account both ecological and economical sustainability. This means to shift *from subsistence to market orientation, from quantity to quality.* Concentration on high quality products (according to local standards!) produced in suitable and ecologically appropriate environments seems feasible and promising for a number of economic reasons:
- a diminished overall importance of agriculture due to increasing numbers of off-farm incomes;
- a slowly, but steadily increasing purchasing power of the inhabitants in Gilgit and other central places both in the Northern Areas and beyond;
- the proximity to markets;
- a positive adjustment to the shrinking availability of human labour-force.

From an ecological perspective, concentration would include the following advantages:
- the abandonment of extremely vulnerable environments such as steep and/or unstable slopes or flood-prone areas;
- the consolidation and rejuvenation of the natural vegetation;
- the overall reduction of human interferences with the fragile ecosystems of these montane environments.

Above all, however, concentration of both agriculture and pastoralism coincides with the urgent necessity to re- and afforestation measures in the highly damaged forest areas of the upper reaches of the Bagrot valley. Reforestation and afforestation would not only contribute to a reconstruction of ecological sustainability, but would also prove to be an employment initiative for the rural population of the Northern Areas at large. This holds true especially if such an initiative would be developed on a large scale with governmental support.

The plea for a shift from subsistence to market orientation and from quantity to quality should not be misunderstood as an attempt to neglect the unquestioned future importance of agriculture and animal husbandry for many households of Bagrot. On the other hand, observations over the last years make it very obvious that socio-economic transformations are underway on a large scale. The increasing dependency of rural households on imported foodstuff is as obvious as the increasing export of quality food products (e.g. seed potatoes!) from Pakistan's Northern Areas. Climatic conditions, location in regard to potential markets, changing lifestyles and consumption patterns, the increasing importance of tourism, growth and competition of non-agricultural income opportunities, growing shortage of manpower or the increasing vulnerability of the fragile ecosystems: these and other factors speak in favour of an economically improved and – at the same time – spatially restricted pastoralism – also in the Bagrot valley.

## 6. REFERENCES

DITTRICH, Chr. (1995): Ernährungssicherung und Entwicklung in Nordpakistan. Nahrungskrisen und Verwundbarkeit im peripheren Hochgebirgsraum. Saarbrücken (Freiburger Studien zur Geographischen Entwicklungsforschung 11).
- (1997): Food Security and Vulnerability to Food Crises in the Northern Areas of Pakistan. In: STELLRECHT, I. & M. WINIGER (eds.): Perspectives on History and Change in the Karakorum, Hindukush, and Himalaya. Köln (Pakistan-German Research Project Culture Area Karakorum, Scientific Studies 3), 23–42.
- (1997): High Mountain Food Systems in Transition – Food Security, Social Vulnerability and Development in Northern Pakistan. In: STELLRECHT, I. & H.G. BOHLE (eds.): Transformation of Social and Economic Relatioships in Northern Pakistan. Köln (Pakistan-German Research Project Culture Area Karakorum, Scientific Studies 5), 231–354.

EHLERS, E. (1995): Die Organisation von Raum und Zeit – Bevölkerungswachstum, Ressourcenmanagemenrt und angepaßte Landnutzung im Bagrot/Karakorum. In: Petermanns Geographische Mitteilungen 139, 105–120.
- (1996): Population Growth, Resource Mangement, and Well-Adapted Land Use in the Bagrot/Karakoram. In: Applied Geography and Development, 48, 7–28.

GRÖTZBACH, E. (1984): Bagrot – Beharrund und Wandel einer peripheren Talschaft im Karakorum. In: Die Erde, 115, 305–321.

HERBERS, H. (1998): Arbeit und Ernährung in Yasin. Aspekte des Produktions-Reproduktions-Zusammenhangs in einem Hochgebirgstal Nordpakistans. Stuttgart (Erdkundliches Wissen 123).

KUHNEN, F. (1992): Sustainability, Regional Development and Marginal Locations. In: Applied Geography and Development, 39, 101–105.

PILARDEAUX, B. (1995): Innovation und Entwicklung in Nordpakistan. Über die Rolle von exogenen Agrarinnovationen im Entwicklungsprozeß einer peripheren Hochgebirgsregion. Saarbrücken (Freiburger Studien zur Geographischen Entwicklungsforschung 7).
- (1997): Agrarian Transformation in Northern Pakistan and the Political Economy of Highland-Lowland-Interaction. In: STELLRECHT, I. – M. WINIGER (eds.): Perspectives on History and Change in the Karakorum, Hindukush, and Himalaya. Köln (Pakistan-German Research Project Culture Area Karakorum, Scientific Studies 3), 43–57.

PRICE, M.F. & M. THOMPSON (1997): The Complex Life: Human Land Uses in Mountain Ecosystems. In: Global Ecology and Biogeography Letters 6, 77–90.

SCHNEID, M. (1997): Identity, Power, and Recollection: Inside and Outside Perspectives on the History of Bagrot, Northern Pakistan. In: STELLRECHT, I. (ed.): The Past in the Present. Horizons of Remembering in the Pakistan Himalaya. Köln (Pakistan-German Research Project Culture Area Karakorum, Scientific Studies 2), 83–110.

SNOY, P. (1975): Bagrot. Eine dardische Landschaft im Karakorum. Graz (Bergvölker im Hindukusch und Karakorum 2).

STÖBER, G. (1998): Zur Transformation bäuerlicher Hauswirtschaft in Yasin (Northern Areas, Pakistan). Unpublished manuscript (to be published in Bonner Geographische Abhandlungen, in preparation).

WHITEMAN, P.T.S. (1985): Mountain Oases: A Technical Report of Agricultural Studies (1982–1984) in Gilgit District, Northern Areas, Pakistan. Gilgit (FAO/UNDP – PAK 80/009).

YOUNAS, M. (1997): Rangelands and Animal Produktion: Constraints and Options. Desertification Control Bulletin No. 31, 30–36.

# LIVESTOCK ECONOMY IN HUNZA
# SOCIETAL TRANSFORMATION AND PASTORAL PRACTICES

Hermann Kreutzmann

## 1. INTRODUCTION

In the discussion of sustainable resource utilization in high mountain regions the time-space relationship of seasonal mobility between locations of favourable conditions plays the dominant role irrespective of focusing on mountain nomadism, transhumance and/or combined mountain agriculture. In a wider context the amendment of such utilization patterns is investigated when change over time is addressed. The prime factor of interest seems to be the degradation of natural resources as an influential triggering parameter, followed by population growth and modernization. The latter two are frequently linked to an opening-up of formerly remote and peripheral regions. Such an incorporation into a wider exchange system is generally linked to a decline of pastoralism and a reduced importance of high mountain agriculture in itself.

The same view holds true for this study area and the advent of modernization has been dated and connected with the inauguration of the Karakoram Highway (KKH) by the end of the 1970s.[1] Earlier transformations are often neglected and underestimated as well as the constant dynamics in high mountain societies which are reflected in all ways of life including the utilization of natural resources such as high pastures. The Hunza Valley has frequently been projected as an example where the impact of the Karakoram Highway is strongly felt, or in comparison to other regions along the course of the KKH where the "winds of change" blow stronger than anywhere else.[2] Consequently the income composition of a Hunza household (Fig. 1) reflects opportunities of livelihood generation from different realms: agrarian and non-agrarian. The agricultural sector itself is structured by two interrelated complexes composed of irrigated crop-farming and animal husbandry. Both contribute to the subsistence of the household as well as they produce marketable goods. The degree and quality of such contributions depends on a number of factors which are directly and indirectly linked to the exogeneous resources. A pattern where non-agrarian income generation plays a major role is quite uibiquitous in high mountain regions. The diversification of resources and the allocation of income from a number of activities has been

---

[1]  For such a perspective see for example Allan 1987, 1989, 1991, Grötzbach & Stadel 1997, Uhlig 1995.
[2]  Mills 1996. Cf. Kreutzmann 1989, 1991, 1993, 1995, 1996, 1998.

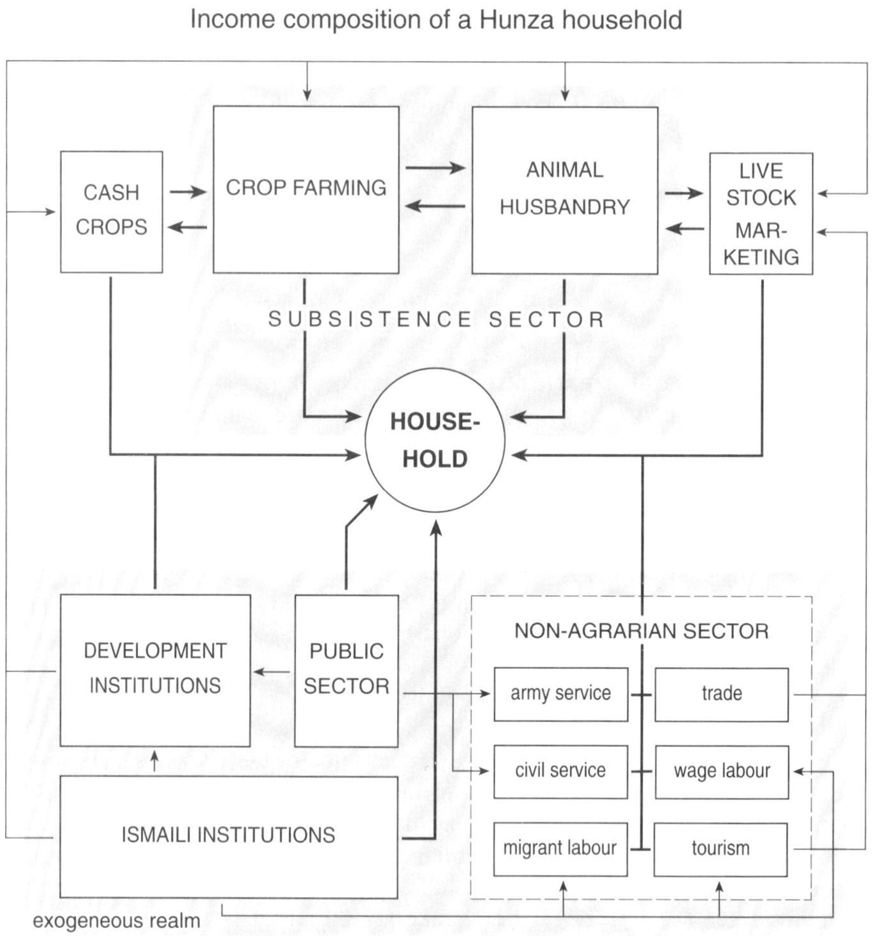

## Income composition of a Hunza household

Fig. 1: Income composition of a Hunza household

present over long periods although very often the superficial impression of a sole dependence on combined mountain agriculture governs the perception of mountain societies. The present structure results from societal transformations over time and an altered set of opportunities for livelihood strategies. The political frame conditions have undergone tremendous changes since the 19th century. Hunza as a territorial entity expanded under the rule of Mir Silum Khan III (1790-1824) northwards across the passes of the Karakoram into present-day China where mainly high mountain pastures were utilized in order to increase the revenue of the Hunza State. This practice was terminated in 1937 when the principality had already experienced 45 years of domination as a quasi-autonomous mirdom within British India. About a quarter century after Pakistan became independent the absolute authority of the Hunza ruler in internal affairs was abolished by the central government and the former state became an administrative subdivison within the Gilgit District of the Northern Areas. Besides changes in political conditions and traffic infrastructure socio-economic transformations affected agriculture and the livestock sector in particular. Understanding high mountain economy as a complex phenomenon and an interrelated set of activities significant changes should be realized in all elements including the livestock sector. Consequently it seems justified to address socio-economic transformations from the perspective of pastoral practices in a given ecological and politico-economical environment. The latter aspect seems to be extremely important as political power structures and transformations affected the local and regional economics significantly.

On first sight the "normal" pattern of the utilization of high pastures is observed in the Hunza Valley as well (Fig. 2). Seasonal migrations take place between permanent homesteads in the arid valley grounds and natural high pastures in the vicinity of glaciers (Photo 1). Their schedule reveals a time-space relationship in the utilization of locally available resources. Beyond the aspect of "what we can see" there occurr a number of questions about the underlying man-made rules and regulations, access rights and livestock productivity, workforce and composition of herds, commonalities and disputes. While focusing on pastoral practices it is attempted to assess the dimensions of change from three perspectives:

– First the importance of animal husbandry for the generation of state revenue and the active role played by the Mir of Hunza in setting the stage for pastoralism within his sphere of influence.
– Second the workforce availability and the division of labour have undergone substantial changes over time and their impact on pastoral practices is strongly felt.
– Third the contribution of the livestock sector for present-day income generation and its role in the dispute about the commons.

Thus it might be possible to shed some light on the transformations having occurred in time, space and quality. Neither nomadism nor combined mixed mountain agriculture appear to be archaic surviving practices. Both are undergoing regular changes, modifications and adjustments. Especially the dynamic adaptation to a transforming socio-political environment and the powerful incorporation into a supraregional market structure needs attention when in the discussion of sustainable development the search for a role model is too often orientated along ecological conditions alone. The neglect of economy and society contributes to the presentation of a somehow distorted representation of pastoral practices. The attention towards animal husbandry and its role in high mountain agriculture is rather challenging as the research about this sector seems in need for more complex approaches and historical depth.

## 2. THE HUNZA STATE AND THE DEVELOPMENT OF REVENUE FROM LIVESTOCK PRODUCTION

A century ago the Hunza population had reached about one fifth of the 46 000 which were returned during the Census of 1998. For earlier periods trustworthy estimates are not available. Oral traditions claim that the population was even lower at the beginning of the 19th century. Only Central Hunza was inhabited by Burusho people in fortified villages (*khan*) while a few Shina speaking settlers occupied the lower part of the valley. The upper Hunza Valley which is now called Gojal was devoid of any permanent dwellings. Seasonally – meaning every summer – Kirghiz nomads crossed the northern passes of Hunza and utilized the high pastures in Chupursan, Mintaka, Kilik and Khunjerab probably even further down the valley (cf. Fig. 2). Grazing taxes (App. 1) were delivered annually to the Mir residing in Baltit (present-day Karimabad) or extracted by his collectors on the grazing grounds. Pastoral practices were divided between Kirghiz nomadic use in the upper part (Gojal) and combined mountain agriculture in Central Hunza and the lower parts of the valley (Shinaki) while it seems to be unlikely that nomads and transhumant shepherds were visiting the Hunza Valley from the South. The extra revenue from affluent nomadic communities was bitterly needed for the upkeep of the frugal lifestyle of the Hunza ruler. Mir Silum Khan III succeeded to expand his territory further north in order to control Kirghiz pastures to a greater extent. Per forty sheep and per thirty yaks a contribution of one animal had been agreed upon as annual grazing tax.

Relations between the nomads of the Taghdumbash Pamir and the Mir of Hunza continued to last for about one and a half centuries (cf. App. 1). They contributed substantially to his income while he sent his own flocks with Hunza shepherds for summer grazing to the Pamirs as well. Mir Silum Khan III is still regarded as one of the most innovative rulers. He initiated the construction of a number of irrigation channels and increased the agricultural lands of Hunza significantly. The internal agrarian colonization characterizes the 19th century and coincides with a shift of pastoral practices within Hunza. The juvenile irrigated oases were meant to support a growing population and to provide shelter for refugees from neighbouring communities, thus increasing the population and – most important – the revenue of Hunza.

In the aftermath a two-fold strategy was applied: The far-away and difficult-to-be-controlled Pamir pastures were allocated to nomadic communities such as the Kirghiz and to peasants dwelling in adjacent Wakhan and Sariqol who utilized those summer pastures in their practice of combined mountain agriculture. In the Hunza valley itself and its major tributaries the ruler tried to settle agriculturists on a permanent basis.[3] Besides some fortified villages with Burusho settlers from Central Hunza mainly Wakhi refugees occupied the single-cropping region in Gojal and colonized the majority of oases at the upper limits of cultivation. As a rule the share of animal husbandry in their combined mixed

---

[3] For a detailed study of the settlement process in Hunza cf. KREUTZMANN 1994.

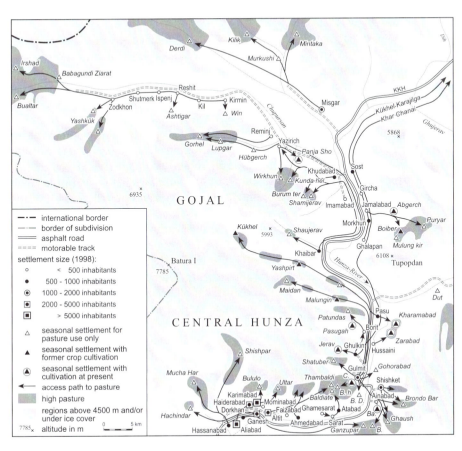
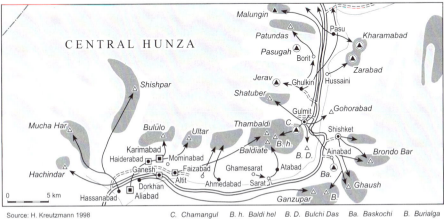

Fig. 2: Settlements, pastures and seasonal migrations in Hunza

mountain agriculture was higher than in the lower-lying parts of Hunza where double-cropping was feasible.

As a consequence of this development Kirghiz nomads lost their traditional grazing grounds and were forced to shift their flocks northwards. Towards the end of the 19th century British intelligence reports claim that the competition between Kirghiz nomads, Wakhi and Sariqoli on the one hand and Hunzukuts with their own and their ruler's flocks on the other led to disputes which were settled by the Chinese representative in Tashkurghan in the Mir of Hunza's favour (App. 1). While in the preceding period the grazing dues were levied in live animals the production of livestock in Hunza itself seems to have increased substantially. From then on the grazing dues (*khiraj*) levied by the Mir of Hunza in the Pamirs were paid predominantly in kind such as felts, woollen blankets (*namdā*), ropes from yak hair, cotton cloth (*kirpas*), coarse cloth (*kham*), coats (*śuqá*), saddles (*jhūl*), rugs (*śarmá*) and socks (*paipakh*). Since the beginning of the 20th century regular payment of grazing dues was received in Hunza which amounted in 1931 to 65 felts, 50 ropes, 25 rolls of cotton cloth and 2 coats. These goods had an exchange value in the Gilgit and Kashgar bazaars and created a welcome extra source of revenue for the ruler. In addition the political influence in the border areas and the participation in Central Asian trade as a transit region enhanced the value of authoritative presence in the Pamirs.

Livestock production in the newly developed oases within Upper Hunza had substantially grown during the 19th century. Gojal provided four fifths of all taxes in Hunza although barely one fifth of the population was settled here. In 1894 Gojal delivered 350 sheep and goats (*kla*) in addition to fifteen *maund* (1 md = 37.32 kg) grain to the tax collectors.[4] This covered basically the regular demand for the Court while additional livestock was available in the personal flocks. The then ruling Mir M. Nazim Khan (1892–1938) backed by the British authorities in Gilgit managed to increase his personal income manifold. The example of the village of Shimshal where in the adjacent *pamér* the best grazing of Hunza is available provides ample evidence. While in 1894 the livestock dues (*ilban*) amounted to 50 sheep (equivalent of 200 Rs), in 1938 the tax collectors extracted a total of 40 md salt (= 400 Rs), 7 md wheat, 16 md barley, 146 sheep and goats, and two yaks. The value of all goods amounted to 1598 Rs which equals an eightfold increase during Mir M. Nazim Khan's reign.[5] Thus the 48 households (app. 400 persons) of Shimshal (App. 2) contributed in 1938 nearly an equal amount of taxes to the Hunza revenue as the whole principality did in 1894 when a total of 1800 Rs was collected.[6] The importance of Gojal in poviding revenue to

---

4   IOR/2/1079/251. Quotations from files and books archived in the India Office Library are referred to under the abbreviations IOR (for records) and IOL (for library) accompanied by the file number: "Transcripts/Translations of Crown-copyright records in the India Office Records appear by permission of the Controller of Her Majesty's Stationery Office."
5   IOL/P&S/12/3292.
6   IOR/2/1079/251. British India and the Maharaja of Kashmir supported the Hunza ruler in 1896 with subsidies worth 3000 Rs annually (GODFREY 1898: 79). For the present state of pasture utilization in Shimshal cf. BUTZ 1996, ITURRIZAGA 1997.

the state is obvious when we register that all livestock of 350 animals per year in 1891 was extracted from there.[7] At that time the whole population of Hunza ranged around 10 000 persons. The resettlement of Wakhi refugees and Burusho colonizers in Gojal and the subsequent severe taxation practices considerably increased the wealth of the ruling family. Emphasis on a controlled grazing policy with higher returns from an intensified animal husbandry were reflected in the stocking density all over Hunza. On a reconaissance tour Colonel R. C. F. Schomberg visited Gojal and acknowledged the overall importance of animal husbandry in the combined mountain agriculture. He regarded the Wakhi pastoral practices within their combined mountain agriculture as equivalent to those of nomads or recently settled nomads.[8]

The exploitative taxation at the climax of Mir M. Nazim Khan's reign coincides with the loss of all rights in the Taghdumbash Pamir, such as the levying of grazing taxes as well as sending flocks from Hunza there. The loss was estimated equivalent to a meagre 200–300 Rs annually and the Mir was compensated by the British authorities (cf. App. 1). While the Pamir revenue had been stagnating for about three decades the livestock taxes within Hunza had grown substantially as had the utilization of grazing grounds.

From a legal aspect the Mir of Hunza regarded himself as the sole proprietor of all natural resources including pastures as a document from 1935 explains: "All forests, mountains and pasture lands in Hunza belong to the Mir and have been granted by him to the different communities who take their flocks for grazing to their respective grazing grounds. The Mir is entitled to graze his flocks in any pasture he wishes. The Mir's wood cutters are at liberty to cut wood from every forest. If the Mir betakes himself on a pleasure trip to any of the above pastures, every shepherd should present him with a sheep and one roghan [wakh. *rúγun* = piece of dehydrated butter]."[9] The hereditary ruler executed some examples of this when he expropriated the grazing rights at Ultar (above Karimabad) for his own herds and when he compensated himself for the loss of grazing rights in the Chinese-controlled Pamir by shifting his personal herds to grazing grounds in bordering Kilik, Mintaka and Khunjerab which were previously used by Hunzukuts' peasants.[10]

---

7 IOL/P&S/12/3292: 156; IOR/2/1079/251: 8.
8 SCHOMBERG 1934: 211; 1935: 169. The nomadic connotation probably stems from a military report (General Staff India 1929: 144) and has been afterwards repeated by geographers such as ALLAN (1989: 135), DICHTER (1967: 45) and E. STALEY (1966: 322). The observed activities very well conform with what has been termed as combined mountain agriculture (in the introductory chapter of this volume) and are quite different from nomadism. Similarly GLADNEY (1991: 37) terms the Wakhi and Sariqoli across the border as "Tajik nomads of the Pamir mountains in southwestern Xinjiang". All of them neglect the significance of irrigated crop farming in the combined high mountain agriculture of these peasants which has been an integral part of their agriculture since settlement there and not merely a recent development.
9 Quotation from an untitled and undated (app. 1935) file from the Commissioner's office Gilgit about the property rights of the Mir of Hunza.
10 LORIMER 1935–1938, II: 259.

It did not escape British intelligence that the exploitation of the Hunza people had reached an undesirable peak. Colonel Schomberg who admired Mir M. Nazim Khan as "... a personality seldom met with in the East. He is a thorough Oriental in every respect, and that is to his credit"[11] observed during his mission in 1934 that the Hunza "... population is little better than serfs. Everything is done at the Mir's orders." The future expectation and a relief option for the concerned population is expressed by Schomberg as follows: "They only ask to be governed as they were before the British came. They ask that their Mirs should rule in future as their ancestors did in the past. The customary laws, especially in Hunza, amply safeguard the rights of the subject ... The Mir is the irresponsible arbiter and autocrat, governing solely for his own advantage ...".[12] The colonial administration in Gilgit feared social unrest at the northern frontier and registered a growing number of Hunza people escaping the state who attempted to settle in the vicinity of Gilgit Town where cultivable lands and jobs were available.

Times had changed since the Gilgit Agency was leased by the Maharaja of Kashmir to British India in 1935. The activities of colonial authorites were directed in improving the "pedigree stock" and in encouraging local farmers to produce more wool in order to meet the demands of the Gilgit Bazaar. At the same time the first ever cattle show was held in Gilgit and merino rams were introduced.[13] For some observers this event marks the beginning of development activities in the Northern Areas and it coincides with the peak of livestock-keeping in Hunza.

## 3. IMPORTANCE AND QUANTITATIVE DECLINE OF ANIMAL HUSBANDRY IN HUNZA

In the same year when revenue from pastoral practices had reached its climax the local historian Qudratullah Beg compiled a survey of all pastures in Central Hunza and Shinaki (Tab. 1). His work augmented by other contemporary sources[14] provides us with a reference which forms the base of comparison with the present situation. Although the number of households doubled in Central Hunza and Shinaki, and quadrupled in Gojal since 1931 the number of persons involved in animal husbandry has decreased substantially. In several cases no shepherds at all are recorded today, in other summer settlements the workforce has been reduced to nominal representation. High pasturing as part of combined mountain agriculture has undergone a significant transformation within the last fifty years. In Central Hunza in 1935 every fifth to tenth household usually sent a shepherd to the high pastures, and in Gojal more than three quarters of all households

---

11 SCHOMBERG 1935: 119.
12 All quotations are taken from Schomberg's confidential report on the social condition in Hunza (IOL/P&S/12/3293: 3–7).
13 Cf. IOL/P&S/12/3294; IOL/P&S/12/3288: Administration Report for 1936.
14 Cf. LORIMER 1935–1938, SCHOMBERG 1934, 1935, 1936, VISSER-HOOFT 1935.

participated in the *kuč* (seasonal migration).[15] Half a century later fieldwork revealed a completely different picture (Tab. 1): little more than one per cent of the households provided shepherds in Shinaki and Central Hunza. Even in Gojal, the number decreased considerably with the exception of Shimshal where the proportion of households following the difficult tracks to the remote pastures resembled the pattern from fifty years ago.[16]

The general impression is that the utilization of seasonal settlements for cultivation and animal husbandry has been reduced, in some cases towards insignificance. The crop farming as a side activity of shepherds in the summer settlements has been given up almost completely. In a few rare cases the cultivated terraces are used for fodder production (grass and alfalfa). The animal mix has remained similar, only horses are not kept anymore. The size of flocks must have diminished quite a bit although comparative data are difficult to come by. Especially in Central Hunza one or two shepherds are nowadyas sufficient to control the herds of their respective communities. With a population of 5000 plus inhabitants in 1992 the four clans of Karimabad accounted for livestock composed of 2224 sheep and goats, and 970 cattle.[17] The average household (size 8.5 members) calls less than four sheep and goats and less than two cows their own. In comparison the mean livestock property in Pasu accounted for twenty-three sheep and goats, and more than seven cattle (including three yaks) per household (8.5 members on average) in 1998. Both villages are far from representative for the overall situation. There is a growing tendency to extend cattle husbandry in the permanent settlements all year round. The cow in the homestead provides the household's need in milk and its derivates.

What are the reasons for such a decline? If taxation was a burden this problem was solved in 1974 when the State was abolished and the fiscal and administrative authorities of the hereditary ruler were terminated. In fact presently the local, regional and national administration does not levy any direct taxes in the Northern Areas at all. The previous heavy-felt burden of livestock and other agricultural taxes was relieved from the farmers and could have resulted in boosting agricultural production. We observed quite a contrasting development and the fate of animal husbandry needs to be discussed in a wider context. The decline of animal husbandry is linked to a general shift in economic activities (cf. Fig. 1). The contributions from off-farm resources have increased significantly. Along with this development a workforce shortage for pastoral activities occurred. Jobs in military and civil services, in trade and tourism require the availability and presence throughout the year. The returns from non-agrarian occupations are higher in general. This holds true especially for out-migrants to Gilgit, Karachi and/or overseas. Even if there are periods with less work load these time frames do not coincide with the heavy burden in agricultural activities during the

---

15 QUDRATULLAH BEG 1935, SCHOMBERG 1936.
16 Cf. BUTZ 1996, KREUTZMANN 1986, 1998, SCHOMBERG 1936: 56, 62.
17 Data according to Karimabad Census 1992 prepared by Karimabad Planning Support Service.

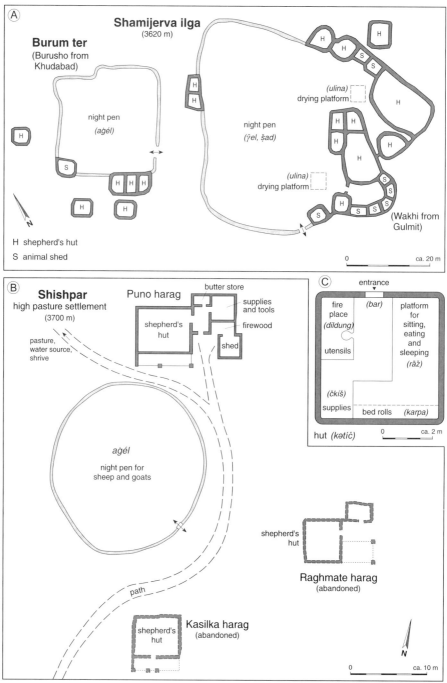

Fig. 3: Locational appearance of pasture settlements

summer season. It becomes more difficult for individual households to spare a member as shepherd. Different strategies are followed to solve the workforce shortage. As off-farm jobs have been predominantly taken up by male household members of a certain age group the agricultural burden was distributed among the remaining household members: women, elderly men and children. The latter age group is nowadays rarely available for agricultural activities in Hunza as almost all children are registered in schools and attend classes. Many of them become young migrants when they continue their education outside their villages. Boys and young men are seen on the high pastures when they accompany trekking groups who very often follow traditional migration paths of livestock or when they are visiting relatives during vacation after a long stay outside of Hunza. Basically the task of shepherds lies presently in the hands of elderly men and women where community rules permit them to go to the pasture settlements. Traditionally this was not allowed among the Burusho and is quite common among the Wakhi. This difference is not revealed from the layout of the combined Wakhi-Burusho pasture settlement in Shamijerav ilga-Burum ter (Fig. 3A). The dual settlement reflects at the same time the two different milk-processing techniques applied in the Hunza Valley (Fig. 4). Wherever livestock is kept in high pastures the production of durable, and highly appreciated consumer goods

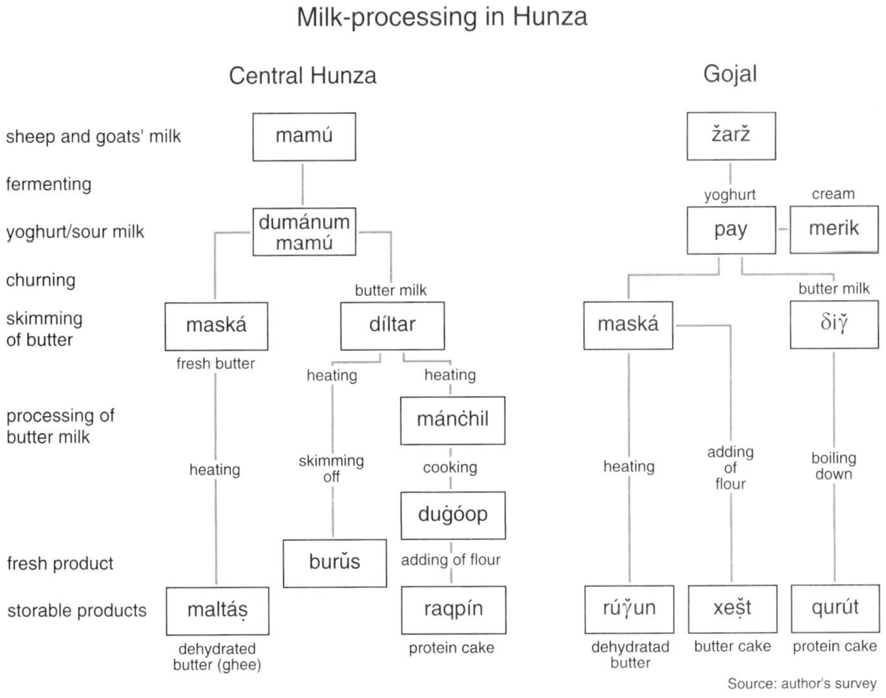

## Milk-processing in Hunza

### Central Hunza

|  | | |
|---|---|---|
| sheep and goats' milk | mamú | |
| fermenting | | |
| yoghurt/sour milk | dumánum mamú | |
| churning | | butter milk |
| skimming of butter | maská | díltar |
| | fresh butter | heating / heating |
| processing of butter milk | heating | skimming off / cooking → mánchil → duġóop |
| fresh product | | burŭs / adding of flour |
| storable products | maltáṣ | raqpín |
| | dehydrated butter (ghee) | protein cake |

### Gojal

|  | | |
|---|---|---|
| sheep and goats' milk | žarž | |
| fermenting | yoghurt / cream | |
| yoghurt/sour milk | pay — merik | |
| churning | | butter milk |
| skimming of butter | maská | δiγ̌ |
| processing of butter milk | heating | adding of flour / boiling down |
| storable products | rúγ̌un | xeṣ̌t / qurút |
| | dehydrated butter | butter cake / protein cake |

Source: author's survey

Fig. 4: Milk-processing in Hunza

Photo 1: Difficult access to high pastures affords sufficient manpower to safeguard the livestock on its way up. This holds true especially for the spring migration when the animals are weak after a meagre winter diet. The passage shown here is on the trail to Shatuber and Bulkish above Gulmit (H. Kreutzmann 19.6.1990).

takes place (Photos 2 & 3). The best part of Wakhi animal husbandry in Gojal is controlled and executed by female household members. Where men are involved the age structure of the shepherds has changed. Elderly men have to take over duties which have been traditionally reserved for the sons of a household for whom it was a privilege to spend the hot summer season in the *Sommerfrische* of the pastures. As the workload is increasing for the permanent household members remaining in the village a tribute has to be paid to agricultural activities.

Photo 2: Milk is processed into butter and the remaining buttermilk is dehydrated by long-time cooking. Sufficient fuel (in the Boiber pasture juniper is burnt) is required to keep the pot boiling for a whole day (H. Kreutzmann 7.7.1990)

Photo 3: The dehydrated buttermilk becomes a durable product (*qurút*) by drying on an elevated platform (*ulina*) in the sun to reduce the moisture content further (H. Kreutzmann 8.7.1990).

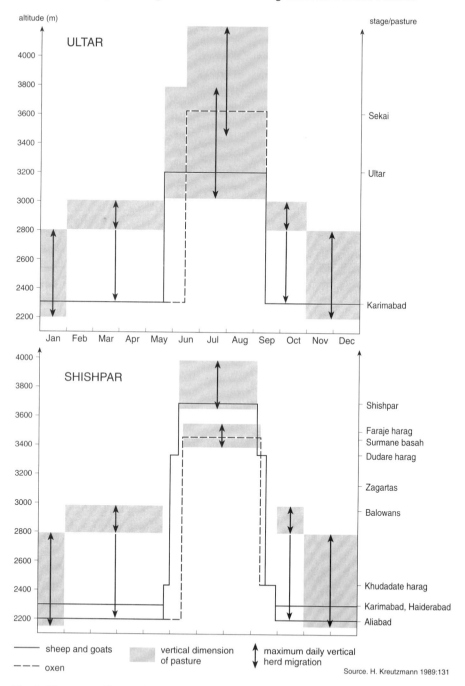

Fig. 5: Time-space diagram for livestock migration in Central Hunza

The flocks have decreased nearly everywhere. No household is presently able to look after sizeable herds all year long. The extra value from animal husbandry is easily compensated if non-agrarian incomes are available. In the village of Gulmit there remained 12 % of all 217 households in 1990 without non-agrarian income, nearly 60 % had more than one additional source.[18] The dependence on animal manure and products has been reduced since mineral fertilizer has been distributed in Hunza. Fresh meat, poultry, milk powder and cooking oil are regularly on offer in the bazaars, even butter fat and *qurút* (dehydrated butter milk) are imported from other valleys. The strong relationship on the agrarian subsistence side in the household economy (cf. Fig. 1) has been weakened by the degree of increased market participation. Furthermore the pasture rotation system (Fig. 5) has been simplified due to lack of sufficient and qualified personnel. Difficult glacier crossings and passages along narrow and steep tracks require the guidance of experienced shepherds (Photo 1). Easily accessible pastures are used more frequently and for extended periods, respectively. This phenomenon leads to exhausting the natural pasture resources in certain areas while giving up additional available pasturage in remote areas. Are we experiencing the final stage of combined mountain agriculture and the end of pastoralism? Does the commercial value of pastures only lie in its fodder, timber and firewood? One aspect needs to be kept in mind. Since the State rule in Hunza was abolished in 1974 the proprietary rights of village lands including pastures have been taken over by clans and village communities. In a region without cadastral surveys and registered property rights the transfer from ubiquitous Miri rule and absolute control over resources to a democratic society with personal landholdings and communal property resembles aspects of agrarian reform. Such a transformation would not necessarily be furnished undisputed. The separation of State and personal property of the hereditary ruler and his relatives from that of the farmers resulted in different perceptions and prolonged negotiations. Immediately valuable resources such as physical infrastructure, communal meeting places, irrigated lands and cultivable waste were affected while pastures and forests fared less prominently in the beginning. This perception has changed in recent years.

## 4. HIGH PASTURES AND THEIR PART IN THE COMMONS

The scenario described above hints at a continous process of lessening importance for animal husbandry and a dramatically changing socio-economic environment. One would expect that these developments would show effects in all walks of life. Do economic activities nowadays take place mainly in the permanent settlements and is entrepreneurship mainly functioning outside the valley? There should be severe consequences for agriculture then. In such a case the

---

18 The 26 households without non-agrarian resources were mainly small households managed by widows or by elderly people without able-bodied members who would be engaged in labour, services or other occupations otherwise. Private enterprise, public services and educational migration plus labour accounted for the dominant occupations.

pasture settlements should look barren and dilapidated. This statement holds true for some places such as Shishpar along the left bank of the Hassanabad glacier in Central Hunza (Fig. 3B). While in 1935 forty to seventy shepherds spent the summers in the three habitations surrounding the animal pen (*aġél*), there number had come down to four in 1985, nowadays mainly two shepherds (*hueltarč*) from Aliabad inhabit the one remaining building while the other two are in ruins (cf. Tab. 1). For certain duties helpers are coming up for the support of the two elderly men, especially when the animals are driven up or down the valley (cf. Fig. 5). On the right bank of the Hassanabad glacier the pastures of Muchu Har are not utilized at all. In Tochi the former pasture settlement has been converted into an orchard (apple, pine trees and willows) and animals are not permitted to trespass. No shepherds at all did visit Muchu Har in 1998.

In other places such as Shamijerav (Burum Ter) above Khudabad the Wakhi from Gulmit have sustained an intact pasture settlement which they share with Burusho from Khudabad who have their own corrall and housing arrangement (cf. Fig. 3A). Both groups utilize their pastures intensively. They sent in 1935 twelve and six shepherds respectively, in the 1990s the numbers were still six and two. Similar developments can be observed in Shimshal where the whole community is actively involved in the summer *kuč* to the *pamér*. Here livestock numbers are still high and since more than one decade fresh yak stock is imported from the neighbouring Chinese Tashkurghan County. The pastures north and south of the Batura glacier (Fig. 6) are utilized by farmers from Pasu and Hussaini. The 87 households of Pasu sent in 1998 a herd of 356 cattle, 282 yaks, 1547 goats and 468 sheep to their Batura pastures, while their neighbours from Hussaini sent only 47 cattle, 677 goats and 214 sheep.[19] The significant change since the previous survey in 1985 occurs in the increase of yak numbers since, all other herds remained quite stable although the number of households (1985: 61) increased.[20] Here we find a pattern where a village community depending to a higher degree on off-farm employment than most other villages sustains a system of pasture utilization through the help of female household members. The productivity of herds has been increased by the introduction of Pamirian yaks.

The labour deficiency has increasingly become a problem for all communities. At the same time attention towards the summer pastures has shifted again. Wherever these pastures are located near the access routes to high mountains or along trekking paths the communities who share the right of pasture claim the right of guiding and portering in these areas as well. They feel entitled to negotiate the terms of trespassing their common property. Villages which are affected by mountain tourism such as Pasu (cf. Fig. 6) and Shimshal have developed a rotation system within their community that all households can participate in this source of income. In a few days more cash income can be allocated to the household than a full summer of shepherding would recover.

---

19 The author is grateful for the collection of data to Einar Eberhardt, Marburg who conducted the Batura livestock survey in 1998.
20 Cf. ABIDI 1987; KREUTZMANN (1986: 102).

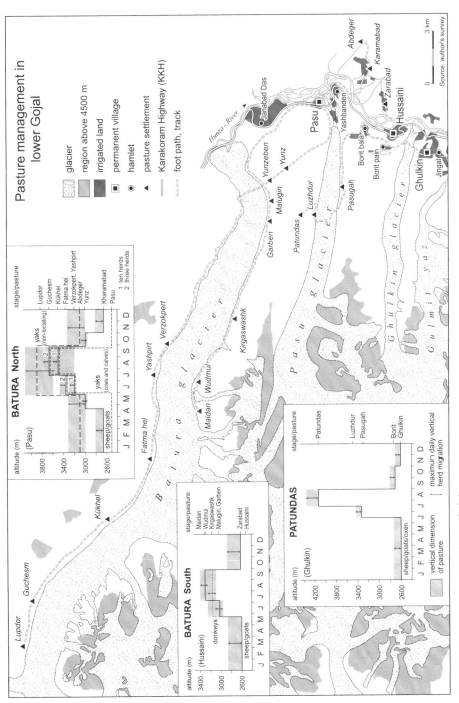

Fig. 6: Pasture management in lower Gojal

Thus mountain tourism has led to a new valuation of high pastures.

A second case is provided by the people of Abgerch, i.e., the Wakhi settlements in the Hunza Valley above Khaiber (cf. Fig. 2). Their traditional right to two pastures in tributaries of the Khunjerab valley – Kükhel and Karajilga – was taken back when the Khunjerab National Park was inaugurated in 1975. The mountain farmers were promised compensation which was not paid until 1990 when a dispute between the Abgerch people and the Government of Pakistan broke out. Negotiations finally led to a settlement which included preferential provision of jobs to Abgerch people in the National Park, control of traffic and hunters, as well as certain access rights to pastures.[21] These mountain farmers faced a dilemma. On the one hand global interest in the protection of Marco Polo sheep affected their pasture resources, on the other hand the protection of extremely old juniper trees in the Boiber valley – their second pasture resource – was supported by the International Union of Conservation (IUCN). The villagers are compensated for not cutting trees there anymore and for restricted pasture use. For these five villages – Ghalapan, Gircha (Sarteez), Morkhun, Jamalabad, Sost – the access rights to common pastures have become a negotiable quantity in their relations with the regional and federal administration as well as with international organizations.

The commons as a village property and one of the last communal resources have gained in importance recently which appears to be contrary to the observations presented above. Never before more village funds have been spent in legal disputes before religious and civil courts (cf. Tab. 1). The village of Gulmit is most severely affected among all and presented here as a virulent case. Gulmit's pastures (cf. Fig. 2 and Tab. 1) are distributed comparatively quite far away from the permanent settlement and not located just above the homestead. During the 1990s different disputes came up with neighbours about the hereditary rights of pasture use. In 1990 a severe dispute began with Shishket across the Hunza river. The Bori kutor clan of Gulmit was to be deprived of its right to access Gaush (Photo 4), the Ruzdor clan made similar experiences in Bulbulkesht and Brondo Bar (cf. Tab. 1). Although blood and marriage relationships exist between the inhabitants of Gulmit and Shishket no solution could be reached through the local institutions and negotiations by mutually accepted and respected neutral persons. The whole conflict escalated and became a principal affair of defending property rights which have not been laid down in written documents. Representatives of public and religious institutions were consulted in vain, before the legal procedures took off. Up to now more than 0.5 million Rs have been spent for lawyers and court fees alone on the side of Gulmit. Similar or even higher contributions were invested by the opponents, not counting all travel expenses and secret meetings of representatives. No solution is in sight besides stay-orders issued by the courts permitting both sides to use the pastures. The funds spent exceed by far the commercial value of these pastures for the next decade. Another dispute

---

21 For a detailed account of this dispute cf. KNUDSEN 1996, KREUTZMANN 1995b, ZAIGHAM KHAN 1996.

Photo 4: The men of Gulmit are marching towards Shishket as the land dispute about the pastures (on the slopes in the left background) seems to escalate (29.4.1990). For almost a decade now funds and resources were invested in this and similar disputes (H. Kreutzmann).

between these two villages occurred about the waterless scrub area of Bulchi Das. The driving force for allocating so much energy in a land dispute is explained by all opponents mostly as a matter of pride and sometime interpret the case as an ethnic dispute. But at the same time it is the hope for potential mineral wealth to be found in barren land and on pastures and the need to develop irrigated land for future generations and, of course, those lands are the only land resources left. If those are lost for a community, the pressure on land is even higher. A similar conflict followed soon about Baldi hel/Baldiate (cf. Fig. 2), in this case the dispute is between Gulmit and Altit. Again huge funds were spent for a dispute on a pasture which most of the opponents had never visited. The tradition of one side claims that the pasture was divided in three sections: the western belonging to Altit, the central to the Mir of Hunza and the eastern to the Gulmitik. The other side tries to make it a point that already during the reign of Shah Ghazanfar (1824–1865) or Mir Ghazan Khan (1866–1886) the immigrating settlers from Altit and Baltit were allocated Baldi hel as their pastures. Obviously no eyewitness could be presented for this viewpoint. The abolishment of hereditary rule in 1974 created misperceptions and controversial interpretations of customary rights. But here it is the case to drive out one party totally and this party has practiced animal husbandry in Baldi hel until recently. In such cases knowledgable and respected village elders are consulted first, in the second instance religious representatives give their advice before public bodies are adressed. All institutions have failed in finding a solution. Now the ultimate instance is approached by selecting honourable men from both sides who are requested to take an oath by the Holy Kuran and then decide about the property rights. The third case involved the Gulmitik in a dispute with the farmers of Khudabad about Shamijerav ilga, the high pasture above Khudabad (Fig. 3A). The customary priority was given to the Gulmitik, while the Khudabadkuts were tolerated in the same grazing grounds. Now the latter attempt to reach a status quo of equal rights and finally aim at separating the pastures in two shares. Again huge funds have been invested for the legal procedures with only little hope for a mutually accepted solution. These cases have been presented as an example for the need of a settlement in those areas without cadastral surveys. On the other hand it is quite clear that any settlement could generate more disputes and could become a painful affair for concerned parties. Currently pastures lead the list of importance in land disputes. Nevertheless stretches of barren land along roads are coming under dispute in growing numbers.[22] From the intensively used permanent settlements – where land disputes between the former hereditary ruler and local farmers were dominant in the 1970s and 1980s – the controversies have been shifted to the extensively used parts of the village lands: pastures and barren lands. The funds spent on these disputes by far exceed their present commercial value and result in huge economic losses every year. Stating on the one hand a decline of the importance of pastoral practices we observe at the same time a rise in valuation of the land concerned.

---

22  To sum up only briefly the areas of pasture disputes in the 1980s and 1990s besides the ones already mentioned: Murtazabad and Hassanabad about Hachindar; Ganesh and Shishket/Gulmit about Gaush and Ganzupar; Pasu and Hussaini about Kharamabad and Zarabad; Hussaini and Ghulkin about Borit; Ghalapan and Kaiber about Dildung kor; Kil and Kirmin about barren land between the two villages; Misgar and Sost about Bell; Pasu and Shimshal about land between Tupopdan and Dut.

## 5. RECENT CHANGES AND FUTURE PROSPECTS FOR HIGH MOUNTAIN PASTORALISM

The perspectives elaborated on above do not complete the picture of the present state of pastoralism and the estimation of natural grazing grounds. Some observations may hint towards a development which would lead to a further decline in agriculture and in pastoralism at all. But at the same time agriculture is very much alive as an important economic resource. In times when the national economy undergoes difficult periods of crises, when unemployment of skilled people grows and when tourism is dependent on global events and affected by market shifts we observe that the traditional agricultural resource base of the Hunza valley experiences its transformation as well and adjusts at higher productivity levels when possible. New and valuable cash crops such as potatoes, seeds, fodder crops, cherries and other fruits have been introduced and become profitable. Crop farming does not depend anymore on the manure supply from animal husbandry, but households prefer some own livestock for their domestic demands and these animals have to be cared for. In the commercial centre of Hunza, in Karimabad, changes have not only affected the individual growth patterns in irrigated agriculture. The common practice of free grazing (*hetin*) of all livestock in the cultivated lands between the harvest of the second crop in October and the germination of the next crop in March/April has been abolished in 1993. Similar community-based legislation occurred in other villages. As a rule it is enforced by the respective local communities. Private ownership of land was traditionally valid only during the cultivation period when animals were removed to the high pastures. After their return the harvested fields were accessible to everybody's animals until the cultivation period began again in spring. This practice which has been regarded by development agencies[23] as one of the most severe obstacles to improved cultivation techniques and an abridged growth pattern belongs now to the past. The ban is strictly supervised and non-compliance leads to severe penalties.[24] Especially farmers who make experiments with new crops are keen to avoid any losses from livestock predations. Fruit orchards, vegetable and alfalfa plots or seed beds are presently in favour and are devoted to utilize the maximum vegetation period. Consequently, all sheep and goats are banned from other people's fields all year long. This decision was reached in a consensual village forum (*jirga*) and resulted in the search for a solution how to deal with the remaining livestock. The farmers of Karimabad rejuvenated their previously

---

23  The Aga Khan Rural Support Programme (AKRSP) and the FAO/UNDP-sponsored Integrated Rural Development Project (IRDP) advocated such a strategy in the early 1980s. Fencing of irrigated terraces was discussed as one solution, others included the keeping of livestock in stables. Both agencies died not succed with their proposals then (AKRSP 1984, SAUNDERS 1983, 1984).

24  In all villages so-called Falai committees were founded and authorized to impose sanctions for trespassing animals. In Haiderabad the fees amounted in 1998 to 5 Rs per sheep, 15 Rs per goat and 25 Rs per cattle. The collected fine is handed over to the owner of the respective plot.

more or less abandoned pastures. The Buroon clan sent six shepherds in 1935 to Bululo with sheep and goats, in 1985 no shepherd supervised the roaming oxen there which were brought up in spring and taken back in autumn. Since 1998 the Buroon clan has adopted a turn (*galt*) system in the same way as Baltikuts (mainly Qhurukuts) and Dom make use of their pasture in Altikutse sat/Bericho chok. As no household can spare a full-time shepherd, the burden of watching the oxen and milking the few sheep and goats is distributed among all households on a two-day-long shift basis. Every household has to safeguard its participation. Milch cows are rather kept in stables near the homestead. Similar developments took place in Haiderabad and Aliabad, and are under consideration in other villages. Reduced numbers of oxen and the vanishing of horses in Central Hunza have opened up the opportunity to use high pastures differently. Ultar ter (3600 m) above Karimabad is the best example for a combined use of livestock-keeping and mountain tourism. Since a few years the pasture settlement has been extended, a camping ground was opened and food and beverages are supplied for the seasonal trekkers. Abandoned oxen pastures such as Sekai (3600m, cf. Tab. 1) have been stocked for the first time with yaks which have been imported from the Chinese Pamirs.

The question of future prospects of high mountain pastoralism within the context of sustainable development deserves a complex answer, at least for the Hunza valley. The evaluation needs not to search for an overall decline or replacement of high mountain pastoralism or agriculture as such by "modern" enterprises and/or services. The assessment of the fate of high mountain pastoralism within the Hunza society has revealed that socio-economic transformations are reflected in all sectors including the pastoral. Adaptations and modifications are influenced to a higher degree by political and societal developments than by changing environmental conditions in the region where these practices are applied. Thus, sustainability has to account for all available opportunities under a given set of frame conditions. Consequently, pastoral practices and the utilization of grazing grounds will remain to play an economic and security-related role in the Hunza economy. The degree of importance for the generation of household incomes may nevertheless vary quite significantly.

## 6. REFERENCES

### (i) Manuscripts and primary printed sources

General Staff India (1928): Military Report and Gazetteer of the Gilgit Agency and the Independent Territories of Tangir and Darel. Simula.

General Staff India (1929): Military Report on Sinkiang (Chinese Turkestan). Calcutta.

GODFREY, S. H. (1898): Report on the Gilgit Agency and Wazarat and the countries of Chilas, Hunza-Nagar, and Yasin, including Ashkuman, Ghizr, and Koh. 1896–97. Calcutta.

India Office Library & Records: Afghanistan Consulate (1927–1947): IOR/12/50/394.

India Office Library & Records: Crown Representative's Records – Indian State Residencies: Kashmir Residency Files: IOR/2/1075/217.

India Office Library & Records: Crown Representative's Records – Indian States Residencies – Gilgit, Chilas, Hunza and Nagir Files (Confidential): IOR/2/1079/251.
India Office Library & Records: Departmental Papers: Political & Secret Internal Files & Collections 1931–1947: IOL/P&S/12/2336, 2361, 3285, 3288, 3292–3294.
India Office Library & Records: Departmental Papers: Political and Secret Separate (or Subject) Files 1902–1931: IOL/P&S/10/278, 826, 973.
India Office Library & Records: Files relating to Indian states extracted from the Political and Secret Letters from India 1881–1911: IOL/P&S/7/61, 66, 157, 170, 181, 205, 222, 231, 233, 243, 252.
McMahon, A. H. (1898): Report on the Subject of the Claims of the Kanjut Tribe (i.e., the people of Hunza) to Territory beyond the Hindu Kush, i.e., to the Tagdumbash Pamir and the Raskam Valley (IOR/2/1079/253:60–67).
Marshall, W. M. (1913): Note on the History of Hunza Rights on the Taghdumbash Pamir (IOL/P&S/10/278: 216–218).
Qudratullah Beg (1935): Hunza ters. In: Lorimer, D. L. R.: Personal Records. Collection of Notes, communication and material collected during 1923-1924 and 1934-1935, archived in SOAS London: MS 181247.
Schomberg, R. C. F. "Hunza Report" (IOL/P&S/12/3293)
Sykes, P. M. (1915): Report on tour to Yarkand. Kashgar (IOR/2/1076/222).

(ii) Secondary printed sources

Abidi, S. Mehjabeen (1987): Pastures and Livestock Development in Gojal. Gilgit (= AKRSP Rural Science Research Programme Reports 5).
Aga Khan Rural Support Programme (1984): First Annual Review 1983. Gilgit.
Allan, N. J. R. (1987): Ecotechnology and modernisation in Pakistan mountain agriculture: In: Pangtey, Y. P. S. & S. C. Joshi (eds.): Western Himalaya: Environment, problems, and development 2. Nainital, 771–789.
Allan, N. J. R. (1989): Kashgar to Islamabad: the impact of the Karakorum Highway on mountain society and habitat. In: Scottish Geographical Magazine 105 (3), 130–141.
Allan, N. J. R. (1991): From autarky to dependency: society and habitat relations in the South Asian mountain rimland. In: Mountain Research and Development 11 (1), 65–74.
Berger, H. (1998): Die Burushaski-Sprache von Hunza und Nager. Wiesbaden.
Biddulph, J. (1880): Tribes of the Hindoo Koosh. Calcutta (reprint: Graz 1971, Karachi 1977).
Butz, D. (1996): Sustaining indigenous communities: symbolic and instrumental dimensions of pastoral resource use in Shimshal, Northern Pakistan. In: The Canadian Geographer 40 (1), 36–53.
Dichter, D. (1967): The North-West Frontier of West Pakistan. A Study in Regional Geography. Oxford.
Gladney, D. C. (1991): Muslim Chinese. Ethnic Nationalism in the People's Republic (= Harvard East Asian Monographs 149). Cambridge, London.
Gordon, T. E. (1876): The roof of the world, being the narrative of a journey over the high plateau of Tibet to the Russian frontier and the Oxus Sources on Pamir. Edinburgh.
Grötzbach, E. & C. Stadel (1997): Mountain peoples and cultures. In: Messerli, B. & J. Ives (eds.): Mountains of the World. A global priority. New York, London, 17–38.
Iturrizaga, L. (1997): The valley of Shimshal – a geographical portrait of a remote high settlement and its pastures with reference to environmental habitat conditions in the North-West Karakorum (Pakistan). In: GeoJournal 42 (2–3), 308–323.
Knudsen, A. (1996): State intervention and community protest. Nature conservation in Hunza, North Pakistan. In: Bruun, O. & A. Kalland (eds.): Asian perceptions of nature. A critical approach. London, 103–125.

KREUTZMANN, H. (1986): A Note on Yak-keeping in Hunza (Northern Areas of Pakistan). In: Production Pastorale et Société 19, 99–106.
KREUTZMANN, H. (1989): Hunza – Ländliche Entwicklung im Karakorum. (= Abhandlungen - Anthropogeographie. Institut für Geographische Wissenschaften der Freien Universität Berlin 44). Berlin.
KREUTZMANN, H. (1991): The Karakoram Highway: The impact of road construction on mountain societies. In: Modern Asian Studies, 25 (4), 711–736.
KREUTZMANN, H. (1993): Challenge and Response in the Karakoram. Socio-economic transformation in Hunza, Northern Areas, Pakistan. In: Mountain Research and Development 13 (1), 19–39.
KREUTZMANN, H. (1994): Habitat conditions and settlement processes in the Hindukush-Karakoram. In: Petermanns Geographische Mitteilungen 138 (6), 337–356.
KREUTZMANN, H. (1995a): Communication and Cash Crop Production in the Karakorum: Exchange Relations under Transformation. In: STELLRECHT, I. (ed.): Pak-German Workshop: Problems of Comparative High Mountain Research with Regard to the Karakorum, Tübingen October 1992 (= Occasional Papers 2). Tübingen, 100–117.
KREUTZMANN, H. (1995b): Globalization, spatial integration, and sustainable development in Northern Pakistan. In: Mountain Research and Development 15 (3), 213–227.
KREUTZMANN, H. (1996): Ethnizität im Entwicklungsprozeß. Die Wakhi in Hochasien. Berlin.
KREUTZMANN, H. (1998): The Chitral Triangle: Rise and Decline of Trans-montane Central Asian Trade, 1895–1935. In: Asien-Afrika-Lateinamerika 26 (3), 289–327.
LORIMER, D. L. R. (1935–1938): The Burushaski Language. Oslo (= Instituttet for Sammenlignende Kulturforskning, Serie B: Skrifter XXIX–1–3).
MILLS, M. (1996): Winds of Change: Women's traditional work and educational development in Pakora, Ishkoman Tehsil. In: BASHIR, E. & ISRAR-UD-DIN (eds.): Proceedings of the Second International Hindukush Cultural Conference. Karachi, Oxford, New York, 417–426.
SAUNDERS, F. (1983): Karakoram Villages. Gilgit.
SAUNDERS, F. (1984): Integrated Rural Development. Pakistan (Northern Areas). Project Findings and Recommendations. Rome (= Terminal Report Pak/80/009).
SCHOMBERG, R. C. F. (1934): The Yarkhun Valley of upper Chitral. In: The Scottish Geographical Magazine 50, S. 209–212.
SCHOMBERG, R. C. F. (1935): Between the Oxus and the Indus. London (reprint: Lahore 1976).
SCHOMBERG, R. C. F. (1936): Unknown Karakoram. London.
STALEY, E. (1966): Arid Mountain Agriculture in Northern West Pakistan. Lahore.
UHLIG, H. (1995): Persistence and Change in High Mountain Agricultural Systems. In: Mountain Research and Development 15 (3), 199–212.
VISSER-HOOFT, J. (1935): Ethnographie. In: VISSER, P. C. & J. VISSER-HOOFT (ed.): Wissenschaftliche Ergebnisse der Niederländischen Expeditionen in den Karakorum und die angrenzenden Gebiete in den Jahren 1922, 1925 und 1929/30, Vol. 1. Leipzig, 121–155.
ZAIGHAM KHAN (1996): Saviours of the lost park. In: The Herald, September 1996, 142–143.

Tab. 1: Pasture utilization in Hunza 1935 to 1998

| Right of pasture | Pasture name, location | Number of households[1] | | | Number of sheperds | | Type of animal | | Cultivation | | Changes 1998 | Disputes 1998 |
|---|---|---|---|---|---|---|---|---|---|---|---|---|
| | | 1935 | 1985 | 1998 | 1935 | 1985 | 1935 | 1985 | 1935 | 1985 | | |
| **SHINAKI** | | | | | | | | | | | | |
| Shinaki[2] | Baiyes | 137 | 344 | 800 | 12 | | H/P | H/O | xx | – | | |
| Shinaki[2] | Maium bar | 137 | 344 | 800 | 10 | 3–4 | H/O/P | H/O | xx | – | | |
| Shinaki[2] | Rui bar | 137 | 344 | 800 | 10 | | H/O/P | H/O | – | – | | |
| Hindi | Deinger haráay | 170 | 303 | 401 | 3 | | H | | – | – | | |
| Hindi | Chashi haráay | 170 | 303 | 401 | 2 | | H | | – | – | | |
| Hindi | Proni | 170 | 303 | 401 | 4 | | H | | – | – | | |
| Hindi | Phulgi haráay | 170 | 303 | 401 | 3 | | H | | – | – | | |
| Hindi | Hundri | 170 | 303 | 401 | 5 | | H | | – | – | | |
| Hindi | Ghumu ter | 170 | 303 | 401 | 4 | | H | | – | – | | |
| **CENTRAL HUNZA** | | | | | | | | | | | | |
| Murtazabad | Bate khar | 112 | 190 | 267 | 3 | | H | | – | – | | |
| Hassanabad | Hachindar | 47 | 72 | 114 | 5 | | H | H | – | – | a) | |
| Ganesh gir. | Hachindar | 127 | 204 | 225 | 10 | | H | H | – | – | a) | |
| Buróon | Muchu har | 220 | 260 | >320 | 40–70 | 4 | H/O | H/O | xx | – | b) | |
| Dirámitin | Shishpar | 250 | 320 | >400 | 40–70 | 4 | H/O | H/O | xx | – | c) | |
| Dorkhan | Ghumat | 45 | 84 | 93 | 2–3 | 1 | H | H | xx | – | | |
| Ganesh kalan | Ganzupar | 90 | 100 | 132 | 15–20 | 3 | H/O | H/O | – | – | | ++ |
| Buróon[4] | Bululo | 100 | 130 | 160 | 6 | – | H | O | xx | tt | d) | |
| Dirámitin[4] | Hon | 100 | 150 | 200 | 4 | – | H | H | – | – | | |
| Karimabad | Ultar | 380 | 529 | 673 | 5–12 | 3–4 | H/O/P | H/O | xx | – | e) | |
| Vazirkuts | Sekai | 30 | 40 | >40 | 4 | – | O/P | O | – | – | | |
| Karimabad | Bulen | | | | – | – | O | | – | – | | |
| Botkuts | Sucash | | | | 2 | – | H | – | – | – | | |
| Mominabad | Bericho chok | 33 | 47 | 53 | 8 | 2 | H | H | – | – | | |
| Altit | Móintas | 178 | 380 | 367 | 4 | 2 | H | H | xx | tt | | |
| Altit | Talmushi | 178 | 380 | 367 | 4 | | H | H | – | – | | |
| Altit | Khuwate | 178 | 380 | 367 | 4 | | H | H | – | – | | |
| Altit | Tiyash | 178 | 380 | 367 | 8 | | H | H | xx | – | | |
| Altit | Ghundoing | 178 | 380 | 367 | 2 | | H | | – | – | | |
| Ahmedabad | Gurpi | 36 | 70 | 126 | | 3–4 | H | H | xx | tt | | |
| Aisabad | Churd | 14 | 20 | 23 | | 1 | H | H | – | – | | |
| tabad | Baldiate | 32 | 73 | 71 | 10 | 4 | H | H | – | – | f) | ++ |
| **OJAL** | | | | | | | | | | | | |
| hishket, | Ghaush, Brondo Bar, Baltin Bar | 34 | 128 | 176 | | | H | H | | | g) | |
| ulmit[5] | Baldi hel | 96 | 208 | 274 | 8 | 2 | H/O | H/O | – | – | f) | ++ |
| ulmit[5] | Ghaush, Buri alga | 96 | 208 | 274 | 2 | 1 | H | H | xx | tt | g) | ++ |
| ulmit[5] | Brondo Bar | 96 | 208 | 274 | | | H | H | | | | |
| ulmit[5] | Shatuber | 96 | 208 | 274 | – | – | O | O | – | – | | |
| ulmit[6] | Shamijerav | 96 | 208 | 274 | 12 | 6 | H | H/O | – | – | h) | ++ |
| hulkin | Patundas | 41 | 83 | 124 | | | | | – | – | | |
| hulkin | Karachanai | 41 | 83 | 124 | | | | | – | – | | |

| Right of pasture | Pasture name, location | Number of households[1] | | | Number of sheperds | | Type of animal | | Cultivation | | Changes 1998 | Disputes 1998 |
|---|---|---|---|---|---|---|---|---|---|---|---|---|
| | | 1935 | 1985 | 1998 | 1935 | 1985 | 1935 | 1985 | 1935 | 1985 | | |
| Hussaini | Batura (South) | 21 | 50 | 70 | 21 | 6–8 | H/P | H/D | – | – | | |
| Pasu | Batura (North) | 22 | 61 | 87 | 22 | 13 | H/Y/P | H/Y | xx | – | | |
| Shimshal | Shuwart/Shuijerav | 54[7] | 123 | 171 | 42 | 80 | H/Y | H/Y | – | – | | |
| Shimshal | Ghujerab | 54 | 123 | 171 | 12 | 6 | H/Y | H/Y | – | – | | |
| Khudabad[6] | Burum ter | 22 | 70 | 127 | 6 | 4 | H | H | – | – | | ++ |
| Abgerchi[8] | Boiber, Puryar, Mulung Kir | 57 | 153 | 335 | 18 | 10 | H/Y/P | H/Y/D | xx | – | i) | |
| Abgerchi[8] | Karajilga, Kükhel (Khunjerab) | 57 | 153 | 335 | ? | – | H/Y | H/Y | – | – | i) | ++ |

Notes and explanations: H = sheep and goats (bur. *huyés,* wakh. *kla*); P = horses (wakh. *yash*); O = oxen (bur. *har,* wakh. *čat*); D = donkeys; Y = yaks (bur. *bépay,* wakh. *ʒuy̌*) xx = cultivation of grain crops in pasture area; tt = irrigated grasslands (bur. *toq*) ++ = disputes have occurred between neighbouring communities who explicitly claim that their right of access to pasture overrules the one of the other party. Generally the seniority of common law is instrumentalized in these disputes.

1. Number of households entitled to access respective high pasture.
2. Excluding Hindi (Nasirabad).
3. These flocks are pastured by Nagerkuts.
4. In fact these clans from Karimabad are utilizing the right of access.
5. The Gulmitik village population consists of sub-groups organized in clans (*kutor*). According to the *kutor* relationship certain pastures are mainly accessed by certain groups: Bori kutor: Ghaush; Charshambi kutor: Baldi hel; Ruzdor: Baldi hel, Bulbulkeshk, Brondo Bar; Budul kutor: Shamijerav, Kunda hel. All Gulmitik are permitted to graze their oxen in Shatuber and Jerav, and all Gulmitik claim property rights in Bulchi Das.
6. The combined pasture area of Shamijerav (Wakhi: white valley) and Burum ter (Burushaski: white pasture) is jointly used by Wakhi and Burusho in a contiguous settlement with separate dwellings (cf. Fig. 3A).
7. According to Schomberg (1936, 38) and Shipton (1938) 50 households with 160 male members had access to these pastures. An untitled file from the Political Agent's office in Gilgit recorded 48 households in Shimshal in 1938.
8. The term Abgerchi includes all Wakhi settlers of common origin who claim to be the first settlers of Morkhun, Gircha, Sarteez, Sost and Ghalapan who jointly are entitled to the utilization of all pastures of Abgerch, Boiber, Puryar and Mulung Kir in the valley above Morkhun. The data of Gojal originate from fieldwork by the author between 1990 and 1998.
a. The pastures at Hachindar are loosing importance for animal husbandry and are gaining in interest for trekking purposes.
b. The pastures in Muchu Har are divided among three clans (Buróon, Qhurukuts and Baratalin) of Central Hunza. Buróon possess access rights for Mandosh and Bagh. Qhurukuts claim Tochi where no animals are permitted anymore since some years ago tree plantations (pines, willows and apple) have converted this pasture into an orchard. The Baratalin's share is in Gaimaling and Bakhor where a water-mill (*yain*) proves that in former time cultivation and pasturing prevailed. None of it exists anymore.
c. Only two shepherds from Aliabad controlled the herds of oxen, sheep and goats in Shishpar in 1998.
d. After a long period the pastures in Bululo are utilized again by the Buróon clan of Karimabad which has introduced a shift (*galt*) system for shepherds. Each participating

household has to provide shepherd's services on a day-to-day basis which is negotiated each season. This practice has been reintroduced since free grazing in the cultivated lands (*hetin*) is banned in most villages of Central Hunza throughout the year (post-1995).

e  Since 1996 a member of the Wazirkuts has introduced yak-keeping in Ultar. In addition the tourism importance of Ultar has grown significantly. Food and camping services are provided for trekkers.

f  After long periods of reduced summer grazing of herds the dispute between Altitkuts and Gulmitik (mainly Charshambi and Ruzdor kutor) about Baldi hel/Baldiate has rejuvenated the interest in seasonal livestock-keeping in these high pastures where in excess wood is available.

g  Since 1990 a severe dispute exists between the neighbouring villages of Gulmit and Shishket about the pastures in Ghaush, Brondo Bar, Bulchi Das and Bulbulkeshk. Presently Shishket is using Brondo Bar alone (one shepherd) while both villages share Ghaush and Buri alga by sending one shepherd each to the high pastures.

h  Five Gojali households performed the pasture duties in Shamijerav. For the first time the community has hired five shepherds from Chitral and Wakhan as support for the women and children in the settlement. Three Burusho households from Khudabad participated.

i  The Abgerchi people have been utilizing the close-by pastures of Boiber etc. (Fig. 2) much more intensively since they were forced to abandon the Khunjerab valley pastures of Kükhel and Karajilga with the inauguration of Khunjerab National Park in 1975. In 1990 a dispute with the Government of Pakistan about non-compensation for pasture loss led to the reoccupation of the Khunjerab grazing grounds. A partial settlement in recent years enabled them to use the latter and to close off the Boiber valley for some years in order to rehabilitate the pastures and woods there. This project is executed in cooperation with development agencies.

Source: QUDRATULLAH BEG 1935 in LORIMER-Personal Records (SOAS); SCHOMBERG 1936; SHIPTON 1938; fieldwork and interviews by author 1983–1998

## APP. 1: REVENUE OF THE MIR OF HUNZA DERIVED FROM GRAZING TAXES IN THE TAGHDUMBASH PAMIR AND SARIQOL

| Year | namdā (felt, woolen blanket) | ropes | kirpas (cotton cloth) | Śuqá (coat) | Source |
|---|---|---|---|---|---|
| | revenue in kind | | | | |
| 1760–1865 | grazing dues from Kirghiz[1] | | | | IOL/P&S/10/278:216 |
| 1865–1878 | no revenue[2] | | | | IOL/P&S/10/278:216 |
| 1875 | Hunzukuts' looting in Taghdumbash[3], extracting goods from Kirghiz | | | | BIDDULPH 1876:116 |
| 1878–1887 | grazing dues from Kirghiz and Sariqoli[4] | | | | IOL/P&S/10/278:217 |
| 1879–1891 | taxes from Shakshu Pakhpu[5] | | | | General Staff India 1928:85 |
| 1888 | 50 | – | – | – | IOR/2/1079/251:8 |
| 1889 | 100 | – | – | – | IOR/2/1079/251:8 |
| 1890 | 90 | – | – | – | IOR/2/1079/251:8 |
| 1890 | attack of Hunzukuts against Kirghiz nomads in Taghdumbash | | | | IOL/P&S/7/61/52, 53 |
| 1891–1895 | no revenue[6] | | | | IOL/P&S/10/278:216 |
| 1895–1902 | grazing dues of Kirghiz and Sariqoli | | | | IOR/2/1075/217:40[7] |
| 1903 | 22 | – | – | – | IOL/P&S/7/157/1277 |
| 1904 | grazing dues of Kirghiz and Sariqoli | | | | IOL/P&S/7/170/1893 |
| 1905 | 33 | 31 | – | – | IOL/P&S/7/181/1554 |
| 1906 | 45 | 30 | – | – | IOL/P&S/7/205/1602 |
| 1907 | 40 | 30 (= total value 200 Rs) | | | IOL/P&S/7/205/1602 |
| 1908 | 40 | 30 | – | – | IOL/P&S/7/222/2027 |
| 1909 | 41 | 35 | 2 kham[8] | – | IOL/P&S/7/233/1556 |
| 1910 | 50 | – | 2 kham | | IOL/P&S/7/243/1407 |
| 1911 | 40 | 40 | 10 | – | IOL/P&S/7/252/1654 |
| 1912–1914 | 40 | 30 | 3–4 | – | IOL/P&S/10/826:173[9] |
| 1914 | 70 | 50 | 9 | – | IOL/P&S/10/826:173 |
| 1915 | regular dues from Sariqoli[10] | | | | IOL/P&S/10/826:140 |
| 1916 | regular dues[11] | | | | IOL/P&S/10/826:107 |
| 1917 | 60 | 45 | 15 | 3 | IOL/P&S/10/826:86 |
| 1918 | regular dues | | | | IOL/P&S/10/826:62 |
| 1919 | regular dues | | | | IOL/P&S/10/826:32 |
| 1920 | 62 | 47 | 17 | 5 | IOL/P&S/10/826:9 |
| 1921 | 65 | 32 | 10 | 4 | IOL/P&S/10/973:240 |
| 1922 | 60 | 30 | 6 | – | IOL/P&S/10/973:217 |
| 1923 | 80 | 40 | 20 | 10 | IOL/P&S/10/973:183 |
| 1924 | 60 | 40 | 6 | – | IOL/P&S/10/973:155[12] |
| 1925 | 52 | 40 | 10 | – | IOL/P&S/10/973:129[13] |
| 1926 | 72 | 45 | 4 | – | IOL/P&S/10/973:104 |
| 1927 | 58 | 42 | 12 | – | IOL/P&S/10/973:81 |
| 1928[14] | 55 | 60 | 10 | – | IOL/P&S/10/973:58 |
| 1929 | 55 | 30 | 10 | – | IOL/P&S/10/973:36 |
| 1930 | 61 | 43 | 20 | – | IOL/P&S/10/973:9 |
| 1931 | 65 | 50 | 25 | 2 | IOL/P&S/12/3285:370 |

Livestock Economy in Hunza 117

| Year | revenue in kind | | | | Source |
|---|---|---|---|---|---|
| | *namdā* (felt, wool en blanket) | ropes | *kirpas* (cotton cloth) | *Śuqá* (coat) | |
| 1932 | | regular dues | | | IOL/P&S/12/3285:329 |
| 1933 | | regular dues | | | IOL/P&S/12/3285:329 |
| 1934 | | no dues[15] | | | IOL/P&S/12/3285:284 |
| 1935 | | no dues[18] | | | IOL/P&S/12/3285:284 |
| 1936 | | no dues[18] | | | IOL/P&S/12/3285:284 |
| 1937 | | termination of all tax demands[16] | | | IOL/P&S/12/3285:233 |
| 1948 | | attempt to regain pasture rights in Taghdumbash[17] | | | IOL/P&S/12/2336 |
| since 1985 | | renewed negotiations about property rights[18] | | | |

1  In the aftermath of the conquest of the Taghdumbash Pamir and following the supplanting of the Kirghiz there by the Mir of Hunza's actions to drive them back to Tashkurgan his control of the grazing grounds resulted in annual demands for grazing taxes from all users. Mir Silum Khan III materialized his claim by erecting a stone monument in Dafdar and he presented scalps of Kirghiz to the Chinese representatives in Kashgar and Yarkand (GODFREY 1898: 74, McMAHON 1898: 5). App. 300 Kirghiz were estimated as having lived there before. The annual tax was supposedly fixed at the rate of one sheep per forty sheep and goats as well as one yak per thirty yaks.

2  During the reign of Yakub Beg (1862–1878) in Kashgaria and his occupation of the Pamir region the Mir of Hunza failed to realize any revenue there. The majority of Kirghiz nomads had given up the Taghdumbash Pamir as their grazing grounds during this period and shifted their activities to Aktash. GODFREY (1898: 74) states that until 1880 no taxes were collected. Subsequent to attacks by Hunzukuts around 1866/67 the Kirghiz of Taghdumbash retreated to Tagharma (cf. GORDON 1876: 115). The sedentary inhabitants of Sariqol took their place and utilized the grazing grounds in the Pamir.

3  In 1875 a group of Hunzukuts is reported to have attacked the camps of Kirghiz nomads in Taghdumbash. They took some Kirghiz as hostages and drove horses and a large flock of sheeps away. The Hakim of Sariqol, Hussan Shah, organized a punitive expedition towards Hunza, liberated the captured Kirghiz and pulled down the fort at Misgar (BIDDULPH 1876: 115–116).

4  Around 1880 Mir Ghazan Khan permitted Kirghiz nomads (20 tents) under the leadership of Beg Kuchmumabad (Kuch Muhammad) to utilize the pastures in the Taghdumbash Pamir. The taxes were fixed as one *namdā*, one *śuqá*, some rolls of *kirpas*, one rope or one saddle (*ǰhul*) per tent (*kirgah*). The amount of levies was depending on the size of the flocks. GODFREY (1898: 74) mentions grazing dues to the amount of one sheep, one *namdā*, one rope and one pair of *paipakh* per tent for this period. Since 1883 annually Sariqoli shepherds visited the same region under same conditions. In 1886 the competition between shepherds from Hunza and Sariqol led to quarrels in Taghdumbash which were pacified by the Chinese representative (Taotai) in Kashgar. According to Mir M. Nazim Khan and *wazīr* Humayun Beg he settled the dispute in a local court in Tashkurgan in favour of the Hunzukuts (IOL/P&S/7/66/701: Letter from J. Manners-Smith to Resident in Kashmir, Gilgit, 4.4.1892; McMAHON 1898: 6).

5  Following a punitive expedition under the leadership of *wazīr* Humayun Beg directed against the inhabitants of Shakshu and Pakhpu (upper Yarkand valley) the annual dues were increased. Until the Hunza Campaign in 1891 the Mir of Hunza received annually five

*yambu* silver equalling 750-800 Rs (GODFREY 1898: 74). The taxes are supposed to having been paid partly in kind: Shakshu: 100 sheep, silver being worth 60 Rs, 2 shot-guns; Pakhpu: 10 *namdā*, silver being worth 60 Rs, 2 shot-guns (according to A. F. Napier in IOR/2/1079/251: Hunza and Nagar Subsidies). During his escape to Yarkand Mir Safdar Ali Khan realized this penalty for the last time. It had been introduced as a making up for the previous enslavement of two Hunzukuts (GODFREY 1898: 74, MCMAHON 1898).

6 In the aftermath of the Hunza Campaign for four years no dues were realized in the Taghdumbash Pamir (GODFREY 1898: 74). This power vacuum was filled by Kirghiz nomads who came back to this region with 200 tents (IOL/P&S/10/278: MARSHALL 1913). Following a directive by the Chinese Amban in Tashkurghan the Kirghiz began again in 1895 to pay taxes to the Mir of Hunza. At the same time the Wakhi settlers of Dafdar (established in 1894) remained exempted from any dues.

7 The taxes amounted to one *namdā* (length of five metres) per household. Impoverished camps would provide smaller *namdā*, one rope or *paipakh* (socks). In addition the shepherds transported all dues as far as Murkushi in Hunza (IOL/P&S/10/278). From 1896 onwards the Mir was supported by the Chinese administration in allocating his dues (MCMAHON 1898: 6). According to the judgement of the Political Agent in Gilgit the revenue of the Mir of Hunza from grazing taxes in Taghdumbash accrued to the value of 200-300 Rs annually around the turn of the century. This amount was higher than the tribute paid by Hunza to China and about a tenth of the value of return gifts received by Hunza from the Chinese Emperor (IOR/2/1075/217:40: Letter from Political Agent Gilgit to Resident in Kashmir, Bunji 6.6.1900).

8 The value of coarse cloth (*kham*) amounted to 2 Rs, that of *namdā* to 150 Rs, that of ropes to 20 Rs the value of one *namdā*-sock (*paipakh*) to 0.5 Rs. During this year Sariqoli (77 households) were the only grazing tax (*khiraj*) payers. They demanded their exemption on the same basis as was applicable for Kirghiz (13 households) and Wakhi (25 households). This resistance could take place because of the little authority displayed by the Amban of Tashkurgan who refused to support the Hunza tax collector Kara Beg. He provided him with 2 *charak āṭā* and 2 *charak* fodder for his horse only while normally a substantially higher provision was allocated: 1 sheep, 1 *charak* rice, 2 *charak āṭā*, 2 *charak* fodder for horses, 2 donkey loads of firewood (IOL/P&S/7/231/1399: Kashgar News-Report 10.8.1909).

9 In 1914 the Wakhi settlers of Dafdar (Sariqol) paid grazing dues for the very first time (IOL/P&S/10/826:173). At this time the grazing community was estimated as composed of 40 Sariqoli, 30 Wakhi and 2-3 Kirghiz households (IOL/P&S/10/278).

10 The Wakhi refused to pay any dues, since the Chinese authorities had threatened them with expulsion if they would not oblige the order to discontinue the tax relationship with Hunza (IOL/P&S/10/826: 155: Gilgit Diary March 1915; IOL/P&S/10/826: 143: Gilgit Diary July 1915; IOL/P&S/10/826: 140: Gilgit Diary August 1915). In the same year the Wakhi of Kilian paid grazing taxes (in currency) to the Chinese authorities for the first time. This agreement was negotiated by the British Consul-General SYKES (1915: 26).

11 The Wakhi of Dafdar renewed the delivery of grazing taxes to Hunza (IOL/P&S/10/826:107).

12 Fewer herds were counted in Sariqol, while an increase was registered for Tagharma.

13 Seven households from Mariang (Sariqol) refrained from accessing the pastures in Taghdumbash Pamir.

14 The exchange value of some items in Kashgar in 1928 (Source: IOR/12/50/394):

| | | | |
|---|---|---|---|
| 1 kg *čaras* | 5–14 Rs | 1 horse | 65–100 Rs |
| 1 kg opium | 40 Rs | 1 donkey | 20 Rs |
| 1 *namdā* | 2,5–3 Rs | 1 sheep/goat | 5 Rs |
| 1 carpet | 25 Rs | | |

15 The riots in Sariqol made the Mir of Hunza decide not to send any tax collectors to Taghdumbash Pamir.

16  In 1937 the Mir of Hunza gave up all traditional grazing rights for his own herds as well as the right to collect taxes in the pastures of the Taghdumbash Pamir. The average profit for Hunza from the Taghdumbash Pamir revenue and the asset from the grazing of a herd of 300 yaks were estimated in the range of app. 200-300 Rs annually during the previous three decades (IOL/P&S/12/3292: 319. The British colonial administration compensated the loss of the Pamirian pastures with an increment of subsidies, the provision of barren land in Oshikandas, and offered the Mir of Hunza potential grazing rights in Naltar Valley (IOR/2/1085/ 296: 23-27).

17  While after the partition of British India the status of Hunza is not clear in respect to its incorporation into Pakistan the Mir of Hunza takes the opportunity to negotiate with the Chinese authorities the option of regaining his traditional grazing rights in the Taghdumbash Pamir (IOL/P&S/12/2361: Letter from E. Shipton to Sec. of GOI, Kashgar 5.4.1948). The Chinese Revolution of 1949 and its expansion into Xinjiang terminates these negotiations.

18  Agricultural reforms in China and in Xinjiang entitled the regional authorities to start talks with the heirs of the Hunza ruling family who claimed property in China (especially in Sariqol and Yarkand) acqired prior to the Chinese Revolution. The negotiations are still in process.

## APP. 2: TAXATION OF SHIMSHAL IN 1938

From Shimshal following tribute is payable to the Mir: From ancient times the people of Shimshal are divided into three sects (1) Dayas [bur. *dāyā* = literally husband of nurse, member of upper class by milk relationship]. From each house of a Daya during the winter 1 he-goat, 1 she-goat, 1 sheep and in place of 1 kid 2 pieces of Shimshali salt, ghee 1 seer [1 *sēr* = 0.933 kg] (it is called ‚Chiragh Tel' [lamp oil]). Puttu 13 yards. During summer 1 he-goat, 1 sheep, 1 sharma [bur. *śarmá*] (carpet made locally), 1 rope, 1 ghirbal [bur. *g̈arbél* = app. 9.3 kg] wheat (10 seers), 2 ghirbal barley and 10 pieces of Shimshali salt. Thus each Daya has to pay annually: 5 goats, 12 pieces salt, one puttu [bur. *páto* = piece of woven woollen material], 1 seer ghee, 1 sharma, 20 seers barley, 10 seers wheat, 1 rope.

From each house of darqans [bur. *tarqán* = literally "without roof, free", in this context a person without the obligation to work as a load-carrier and exempted from certain taxes] during winter, 1 he-goat, 1 she-goat, 1 sheep, 1 kid or two pieces of salt, ghee 1 seer, 1 puttu 13 yards, wheat 10 seers, barley 20 seers. During summer 1 he-goat, 1 she-goat, 1 rope and 10 pieces of salt annually.

From each house of coolies [= people with the obligation to work as load-carriers] 1 she-goat, 1 sheep, in place of kid 2 pieces of salt, barley 20 seers, wheat 10 seers, salt 10 pieces, in place of shoes and socks 2 pieces of salt. During summer 1 he-goat of 2 years, 1 rope annually.

On celebration of marriages of sons and daughters of the Mir from each house of a Daya 1 he-goat and 1 she-goat Rs. 12/- in cash. From Darqan 1 goat or Rs. 8/-. From each house of a cooly Rs. 2/-. At present there are following houses in Shimshal:

| | |
|---|---|
| Dayas | 24 |
| Darqan | 18 |
| Coolies | 6 |
| Total | 48. |

[The Census of 1931 listed 54 households and a population of 398 persons]

The custom is that the grain belonging to the Mir is carried by the people of Shimshal themselves to Gilmit [= Gulmit].

The Trangfa [bur. *tranphá* = village elder] of Shimshal (motabar [pers. *muᶜtabár* = respectable]) who is appointed by the Mir receives annual[l]y from the Mir 4 goats and 2 pieces of salt from each house of Darqan and 1 piece of salt from each house of a cooly. The wool which is called ‚Kachani' [?] is given to the Trangfa. ...

From ancient times Trang[f]as (notables) at Gujhal are appointed from Hunza. Each Trangfa is entitled to goats from three houses of Darqan and three houses of coolies. Arbaba [wakh. *arbāb* = appointed village representative] (lamberdars) are entitled to goats from 2 houses annually. Their wool is called ‚Kachan'. One rope and a small quantity of curd from each house is given to the Trangfas.

Source: Excerpts from an untitled and translated document at the Political Agent's Office in Gilgit dated May 1, 1938 collected by the author in 1998, explanations in brackets are incorporated according to BERGER 1998 and written communication with Prof. Dr. Georg BUDDRUSS, Mainz 1999.

# PASTORAL SYSTEM IN SHIGAR / BALTISTAN. COMMUNAL HERDING MANAGEMENT AND PASTURAGE RIGHTS.

### Matthias E. Schmidt[1]

## 1. INTRODUCTION

Pastoralism involves an interaction between land, livestock and labour, and is in the case of Shigar (Baltistan) confronted with the challenges of the high mountain conditions of the Central Karakorum such as severe winter temperatures, difficult topography, and meagre seasonal fodder resources. The people of the area have managed to utilize different ecological niches within this harsh environment and established their own specific agro-pastoral system. Their choice of which types of livestock to keep and the means by which they can best exploit different altitudinal and seasonal range resources represent adaptive responses to the high altitude conditions of Shigar. However, although environmental adaption certainly shapes pattern of herd composition and herding movements, pastoralism in Shigar is a complex activity that also reflects a number of other cultural, social and economic factors. Intra- and inter-household cooperation in herding practices as well as community regulations of livestock management are important related aspects.

The aim of this contribution is to illustrate indigenous pastoral practices and seasonal herding patterns and to highlight recent developments of the pastoral economy in Shigar. Since investigations about pastoralism in Baltistan are very limited it seems necessary to explain in detail different aspects of livestock keeping in the area.[2] Although statements within this paper concern the Shigar valley in general, the focus of investigations and their findings is based on the example of Shigar Proper.[3] It serves to illustrate certain aspects of herd composition, herding management and practices as well as types of pastures, pasture settlements and implemented pasturage rights. Recent developments connected

---

1 The author is indepted to the German Research Council (Deutsche Forschungsgemeinschaft) for its financial support and to Prof. Dr. Eckart Ehlers for his scientific guidance. Special thanks go to Claudia Polzer for her kind cooperation and to Prof. Dr. Hermann Kreutzmann for his critical comments. As always, the author would like to express his gratitude to the people of Shigar for their hospitality and contributions to this work.
2 MacDonald (1994; 1998a) presents detailed information about animal husbandry in Askole village (Braldo); Hewitt (1991) for Chutrun (Bashe), and Jensen (1997) for the Thale valley, briefly cover pastoral aspects.
3 Fieldwork was carried out between March and November 1997 and in summer 1998, mainly in and around Shigar Proper.

with the construction of new roads and the increasing mountaineering tourism in the area poses the question of how these changes affect the implemented agro-pastoral systems. It should be looked at to what extent the new facilities for education, employment and income, as well as the introduction of farm machinery, improved crop varieties and breeding stock influenced the pastoral structures. The new possibilities and challenges thus not only lead to a change of labour division between households and household members but also of appearance and reputation of pastoral work.

## 2. ECOLOGICAL AND SOCIO-ECONOMIC CONDITIONS

Shigar is situated just north of Baltistan's district capital Skardu in the Central Karakorum. Its major water course, the Shigar River, is formed by the confluence of the Braldo and Bashe Rivers, and is a tributary to the Indus River. The three valleys: Braldo, Bashe and (lower) Shigar are administratively connected to the Shigar subdivision with its central place at Shigar Proper.[4] Whereas Braldo and Bashe, extending roughly from 2,500 m to 3,000 m, are narrow valleys, with villages located on terraces several metres above the rivers, the lower Shigar valley is relatively wide and its villages are situated on alluvial fans. All villages or village clusters form oases in the semi-desert environment.

The combined mountain agriculture still forms the main economic basis for the valley's population. The arable land depends entirely on irrigation. Apart from a few exceptions, land holdings are very small,[5] therefore many households barely meet subsistence needs. Main crops are wheat and barley. A second crop, usually buckwheat or sometimes millet, is possible following barley in regions up to 2,600 m. Main crops of the single cropping zone, the upper Braldo valley, are wheat, buckwheat and peas. Numerous fruit trees, especially apricots in great variety, are grown in gardens (*tshar*)[6] or along irrigation channels of the lower Shigar valley. Gardens are fenced with stone-walls and thorny hedges against sheep and goats while the arable fields are unprotected. Irrigated and stone-walled meadows (*ol*) are situated at the edge of the cultivated area, mostly at steep slopes or at places with poor soil quality. These fields are private property and kept mainly to supply fodder for the livestock. During spring and summer they are irrigated every ten to twelve days. The meadows are rarely manured by dung spreading, but by grazing animals, mainly in the autumn.

In addition to the cultivated areas in the valley floors each village has affiliated pasture areas at different elevations. The most valuable of these com-

---

4 Place names are often used both by the local population and in the literature, in a multi-level and inconsistent way. In the following, Shigar refers to the whole area – Braldo, Bashe and lower Shigar valley. The central place is here called Shigar Proper; it comprises several villages of the two Union Councils Markunja and Marapi.
5 The majority of households possess between 10 and 20 *kanal* land (1 *kanal* = 505,86 m$^2$), i.e. a household rarely owns more than 1 ha.
6 Terms in Balti language are given in italics.

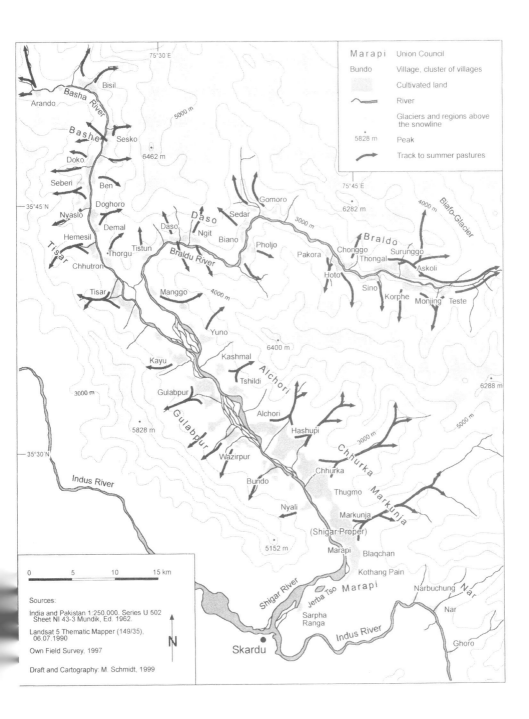

Fig. 1: Shigar, Bashe and Braldo valleys with affiliated summer pastures

mon grazing lands are summer pastures (*bloq*), located on the floors and slopes of the side valleys (s. p. 129) between 3,200 and 4,600 m. Permanent summer grazing camps or pasture settlements used for shelter for men and animals, and for dairy production are situated at altitudes of about 3,200–3,800 m. Fig. 1 shows the main villages of Shigar and the approximate location of the affiliated pasture areas.

Like in other regions of the Karakorum a prominent feature of climate and vegetation patterns in Shigar is the great variation referring to altitude and aspect. A major cause of this is the significant difference in precipitation, humidity and the varying periods of snow-coverage. The vegetation of the lower subalpine areas is influenced by arid to semi-arid conditions, whereas it is primarily humid for the plants of the alpine and sub-nival level. Consequently, Shigar provides pastures at several distinct altitudinal levels.

The heterogeneous moraine and gravel base of the valley floor and lower slopes show no real plant societies while *Artemisia* (*burtse*) is the dominant plant on the extensive, steep valley slopes. These *Artemisia* steppes offer rangeland grazing that is especially useful to sheep and goats in spring. Different shrubs and bushes are predominant along water courses. Juniper trees (*shukpa*), the main source of firewood, occur around 3,000 m in limited number but form closed forests on northern-aspect slopes between 3,300 and 3,900 m. Above 3,600 m there is good summer grazing in the large expanses of subalpine and alpine grassland on the floors and slopes of the wide, glacier-carved upper valleys. The grassy vegetation with a large number of species shows an exposure-dependant differentiation as well with more or less open forb-rich steppes on sun-exposed slopes, while closed sedge mats are confined to shady areas. Small willow bushes (*bloqlhchang*) in particular cover northern slopes between 3,700 and 4,200 m, mainly near or along water courses. In the high alpine and subnival zone at altitudes of about 4,600–4,800 m patchy and sparse alpine vegetation forms final grassy seats that are accessible only in July and August (cf. HARTMANN 1968, 1972; MIEHE et al. 1996: 199–200).

## 3. COMPOSITION OF HERDS

### *Hyaq, Crossbreeds, and Cattle*

Pastoralism in Shigar is mainly based on the herding of different varieties of cattle. A special feature of Balti pastoralism is the cross-breeding. In order to adress this complex aspect of herd composition certain terminological points need to be clarified. The Balti word *hyaq* (yak) refers strictly only to the males of the species *Bos grunniens*, the female yaks being called *hyaqmo*. Similarly, the females of the hybrids which result from the extensive interbreeding of yak with the common domestic cattle (*Bos taurus*) (referred to here as 'common bull' (*xlang*) and 'common cow' (*ba*)) are marked with the suffix -*mo*. The detailed nomenclature for the different generations of hybrids is given in Tab. 1. However, the terms for the F3 and later generations are little used, partly because ancestry of such an animal were not known or remembered by the informant, but mainly because there are fewer of them, resulting from a decline in fertility.

Table 1: Different generations of cattle hybrids in Shigar

|    | Parents | | Offspring | |
|    | (male) | (female) | (male) | (female) |
|----|--------|----------|--------|----------|
|    | hyaq x | hyaqmo   | hyaq   | hyaqmo   |
| F1 | hyaq x | ba       | zo     | zomo     |
| F2 | hyaq x | zomo     | gar    | garmo    |
| F3 | hyaq x | garmo    | gir    | girmo    |
| F4 | hyaq x | girmo    | hluk   | hlukmo   |
| F1 | xlang x | hyaqmo  | zo     | zomo     |
| F2 | xlang x | zomo    | tol    | tolmo    |
| F3 | xlang x | tolmo   | baju   | bajumo   |
|    | hyaq x | tolmo    | tolzo  | tolzomo  |

Source: Own survey 1997; cf. Afridi 1988: 283.

The presently kept *hyaq* can be described as a semi-domesticated form of the wild *hyaq* that is indigenous to the Tibetan plateau. They are noted for their heavy coat that ends in a distinctive fringe of long hair on the legs and flank and are complemented by a bushy tail. *Hyaq* are neither used as beast of burden for carrying loads nor as draught animals for ploughing and threshing purposes as they are in some other high-altitude regions.[7] A reason for the latter can be seen in the fact that *hyaq* are said to be unable to withstand the summer heat of the deep valleys. On the other hand, the common bulls and cows kept in Shigar are small and cannot tolerate the altitude of the upland pastures. Resulting from this, the hybrid husbandry that has a long tradition in the region is practised with the aim to produce animals with more advantageous qualities in regard to fitness, environmental adaption, and milk yield. The most common hybrids, *zo* (female: *zomo*), a mixed breed of *hyaq* and cow (Tab. 1), seem to be particularly adapted to regional conditions. The male *zo*, larger than the common bull but easier to handle than the *hyaq*, is mainly used for ploughing and threshing purposes (Photo 1). As tractors and threshing machines become more and more common in the valley, the necessity of acquiring *zo* is decreasing. They are often slaughtered or sold before reaching maturity. The female, known as *zomo*, is highly valued, being reputed to yield more milk than a *hyaqmo*, but of nearly as high quality. The hair of *zo* and *zomo* is black and short except on the flank and tail where it forms a long shaggy fringe. Features of the tail, horn and fell serve as signs of identification. After four generations the specific features become indistinct so that the farmer cannot give details about the origin of the animal (cf. FRIEDL

---

[7] Cf. KREUTZMANN (1996: 35) for the Pamirs; MANN (1986: 106), OSMASTON et al. (1994: 201) for Ladakh; STEVENS (1993: 149) for Khumbu (Nepal).

Photo 1: Ploughing with *zo* in Shigar (6. 3. 97)

1983a: 71). All the female hybrids are fertile with decreasing fertility from one generation to another while all the males are sterile.[8] Consequently, many farmers have difficulties in maintaining their herd size, let alone producing surplus stock for sale. Since *zo* and *zomo* are preferred, the other types of the above mentioned generations of hybrids are not very common in Shigar.

While common bulls are few, cows are kept by most households. Due to their small size they can also be fed by poorer households with limited winter fodder resources. Cows are valued for their milk, and are important for cross-breeding. Furthermore, the manure of all cattle is regarded as an important resource particularly for fertilizer which is, for some households, the most crucial contribution of livestock to the domestic economy. Additionally, some cattle dung is dried and used as fuel during winters. A common occurrence in autumn is people collecting manure from the stock grazing in the open field area.[9]

---

8   OSMASTON et al. (1994: 207–212) state for Zangskar, a similar environment, that fertility of cattle and hybrids in general is relatively low, with females calving only in alternate years or even less.

9   Dung is rarely carried down from the high pastures as is practised in Ladakh (cf. FRIEDL 1983b: 244).

## Goats and Sheep

Goats (*rabaq*) and sheep (*lu*) have also long been a component of livestock-keeping in Shigar. Both are valued for their hair and wool, their meat, and their manure. Goats are also raised as a main source of milk supply. The goats' hair is cut twice a year (in spring and autumn) and spun on wooden spindles into long strings. The spun strings are used for weaving long and broad strips which after sewn together make a rug (*chari*). Women's caps (*nating*) are also made of goat hair. Sheep are sheared in spring and their wool is used for making blankets (*kar*) or caps. In former times, the people of Shigar weaved rough clothes such as long coats, shawls and footwear by themselves. A special type of carpet is also made of sheep wool by beating and felting it. Usually shearing, spinning and weaving of goat hair is men's work, while the similar work connected with sheep wool is done by women. Like goats, sheep are sacrificed on special occasions like Moharram, Id-e-qurban and other Muslim holidays. Hides of sheep and goats have been manufactured as footwear or leather bags, but are mostly sold to traders nowadays. Goats and sheep are occasionally sold when households are in need of cash.

## Donkeys, Horses, and Poultry

Today, donkeys (*bongbu*) and horses (*hrta, rgunma*) are not very common in Shigar. Since the valley has been connected by a jeep-track with the next major center Skardu, and most goods are carried by jeep or tractor, these load animals have become unnecessary for this purpose. The few donkeys of Shigar are kept for carrying manure to the fields or transporting down dairy products from the high pastures. Today, and probably also in the past, most farmers do this kind of work by themselves without using any load animal. One reason for the small number of donkeys is the lack of fodder to feed them on during winter. They are usually kept at the villages during the summer, so that they are readily available for use. Today, approximately ten to twelve horses are held in Shigar and are only used for playing polo which has always been held in high esteem.

Poultry (*biapho, biango*) is mainly raised for self consumption and fowl birds are only kept in small numbers. Today there are two poultry farms in Shigar that sell chicken and eggs to shopkeepers. In addition, fowl and eggs brought by traders from Skardu, are also sold in the valley. Chicken and eggs are thus rare commodities, but are offered to guests when no other meat is available.

## Herd and Flock Size

Up to this day pastoralism in Shigar is carried out as a subsistence and, only in rare cases, partly as a commercial activity. Mixed stock herds are the common pattern of livestock ownership. Almost all stockowners keep cows or goats as a source of dairy products for household consumption. A household's wealth is traditionally judged primarily by the size of its landholding and the number of livestock, while both are interdependent. Households who own more than thirty head of mixed goats and sheep and as many as half a dozen *zo* or *zomo*, are considered extremely well off. However, excessive stock holdings are limited by the need to provide winter fodder, and only a few households can afford the land and labour required to maintain larger herds.

Tab. 2 shows estimates of the livestock owned by an average household in different parts of the area. Although there are some uncertainties in the figures, some points stand out clearly: The number of livestock kept on average by a household in the different Union Councils reflects roughly the size of pasture land that is at the disposal of the villages. It is highest in upper Braldo and Bashe due to their extended pastures while the lowest rate is given in Gulabpur where there are limited grazing grounds in the steep mountain chain between Shigar and the Indus River. The total pasture area of Shigar Proper is quite large but in connection with the high population number the grazing area per capita or household is relatively small.

Characteristic for large herds is the Bashe valley where much butter is produced by the local farmers. Butter from Bashe has been sold for a long time or exchanged for dried apricots in Shigar Proper. In all parts of Shigar, female cattle are kept in twice as high numbers as male ones. Most households keep less than two adult male cattle, only in Braldo and Bashe is the number significantly higher. The major reason is given by the fact that ploughing and threshing is done in the lower Shigar valley by tractors while it is still carried out with *zo* or bulls in Braldo and Bashe. Goats are raised in larger numbers than sheep since they are not only a major supplier of dairy produce but can cope better with the meagre fodder conditions, especially in spring. Other (modern) reasons for the regional differences in herd size such as the possibility of selling animals to expeditions or the consequences of non-farming employment are discussed later.

Table 2: Herd and flock size per household in Shigar (by Union Councils)

| Union Council | Male bovines (hyaq, zo, bulls) | Female bovines (hyaqmo, zomo, cows) | Goats | Sheep |
|---|---|---|---|---|
| Marapi | 2.3 | 4.6 | 10.5 | 6.2 |
| Markunja | 1.8 | 3.1 | 6.2 | 2.8 |
| Churka | 1.5 | 3.8 | 7.2 | 9.0 |
| Alchori | 2.2 | 6.5 | 8.4 | 8.0 |
| Dassu | 1.6 | 3.0 | 6.1 | 5.9 |
| Braldo | 5.6 | 10.1 | 13.8 | 16.1 |
| Gulabpur | 1.2 | 4.4 | 7.4 | 3.2 |
| Tisar | 2.6 | 6.8 | 6.6 | 5.3 |
| Bashe | 4.0 | 10.6 | 15.2 | 7.2 |

Source: Own survey, 1997–98.

## 4. ORGANISATION OF PASTORALISM IN TIME AND SPACE

*The Annual Cycle of the Agro-Pastoral System*

The organisation of pastoralism in time and space is closely connected with agricultural needs and is dependent on the seasonal availability of fodder for the livestock. For instance, agricultural tasks like ploughing and threshing demand the availability of cattle's traction power at specific times and places. Or to use another example, it is necessary during the growing season to keep livestock away from cultivated areas to avoid damages in the crop land. On the other hand, there are insufficient grazing grounds near the villages and productivity as well as accessibility of different pastures vary during the year which leads to the necessity of a seasonal movement of livestock through an elevational range of pastures to allow the animals access to diverse fodder resources. The migration cycle that may vary from village to village should be illustrated by the example of Shigar Proper (cf. Fig. 1). All high pastures of Shigar Proper are localized in the connected side-valley, called Shigar Nalla by the local population.[10]

The best starting point for tracing the annual migration of herds may be midwinter, because only at this time are all livestock, apart from some *hyaq*, located roughly in similar places – in the stables of the main villages. On sunny days in late winter and early spring, goats are led out to graze on the slopes near the villages where they find *Artemisia*. In March and April, stockholders of Agepa, Astana, Shopa and Ghzoapa take their herds on daily movements to transitional pastures in the Shigar Nalla – Kaniskar, Yulba and Lhchaxmachan (Fig. 2). These privately-owned meadows are very limited and are not sufficient for the livestock of all villages of Shigar Proper. When the fields are ploughed in March, which is done by tractors, or with the help of *zo*, in case of small and badly accessible fields, *zo* are driven to the high elevation pastures (*bloq*) where they graze unattended. Sheep and goats are moved up valley in May. All animals are collected at Kaniskar village-by-village and given to a few men who take them up to the *bloq*. On daily movements the flocks are driven to high elevated meadows following the receding snow line to as high as 4,500 m. The timing of herding in different parts of the alpine pastures may vary, reflecting perceptions of the value of grass in different places. Cattle and *zomo* are moved up to the *bloq* in mid June by the stockholders themselves.

The newly born calves, lambs, and kids as well as one or two lactating animals per household for the daily milk supply are retained in the villages during the whole summer. They are fed with fresh grass and weeds removed from the fields by women. The village community or in some cases the village elderly (*tsharma*) select a village guardian (*lura*) who protects the crops to ensure that no animal enters any field until the fields are harvested. The *lura* usually serves for

---

10  In the vernacular Balti, this side valley is called Shigar Phuchaph, while the Urdu form Shigar Nalla (*nālā*) is more common today. On most maps it is wrongly called Bauma Lungma, a name that is not used by anyone of the area today.

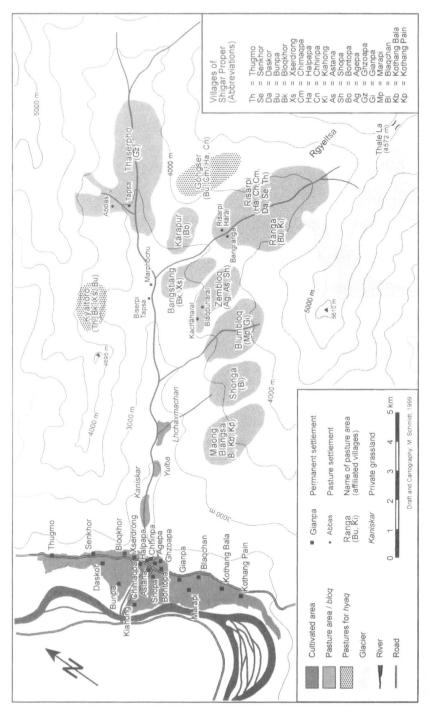

Fig. 2: Villages and affiliated pastures of Shigar Proper

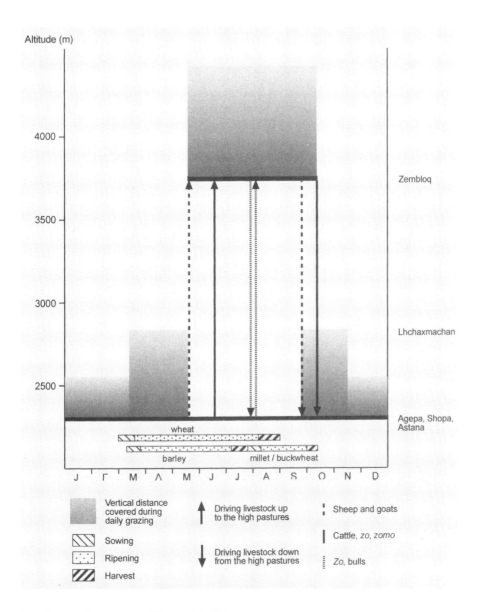

Fig. 3: Annual movement of livestock in Shigar Proper

one year and receives 5 kg wheat from each household as payment for his duty. In addition if he catches grazing animals in the fields, he receives fines from the owner of the specific animal.[11]

11 In Zangla (Ladakh), a watchman, called *lorapa*, performs a similar practice. Besides taking care that no animal enters the fields he tends all the livestock of the village that stay in the valley during summer, and receives wages per tended animal (FRIEDL 1983b: 244).

At the end of August, some *zo* are driven down from the *bloq* for ploughing the fields after the barley harvest. Most of them are brought up to the high pastures again after ploughing. All the livestock that have been taken to the *bloq* spend the rest of the summer up valley until after the crops have been harvested. From mid-September on, some of the valley-based stock – cows, young sheep and goats – can be seen in increasing numbers grazing in the cropped fields. Cattle are tied while sheep and goats graze freely. In the middle of the month the grass of the irrigated meadows (*ol*) is cut by men and women with a sickle. Once cut, the grass is spread on the field for several days to dry and is then carried to the houses where it is stored as winter fodder. Households normally cut meadows near their homes themselves, but it is common to help each other for more distant or larger fields.

Sheep and goats are driven down from the pastures by a few men at the end of September and fetched by the owners at Kaniskar. The livestock of Agepa, Astana, Shopa and Ghzoapa graze the meadows of Kaniskar, Yulba, and Lhchaxmachan for two months after the grass is cut. Cattle, *zo* and *zomo* are brought back down from the *bloq* at the beginning of October (Fig. 3). From then on, the livestock is allowed to graze freely on stubble in the harvested fields, which contributes to field fertility as well as nourishing the livestock. As the weather grows colder the stock is stabled overnight in the lower story of the traditional two-story house. On warmer days, goats and sheep are led out to graze adjacent rangeland – the nearby slopes or the flat grasslands near the Shigar River (*tsosa*) – but are kept in stables during the night. Today, hardly any family in Shigar Proper lives together with its livestock in the traditional winter room (*kazah*). Most people nowadays prefer to spend the winter separated from their stock in the upper floor of their house. The most recent tendency is the construction of single-storey houses with a separate building for the livestock.

## *Winter Feeding the Livestock*

During the summer, the livestock is sent to high elevated grazing grounds where the animals have rich alpine pastures at their disposal. Due to the harsh climate and the poor and limited winter pastures the livestock must be fed fodder in order to survive the winter and early spring. The task of supplying fodder for the livestock during that time is one of the major problems for the local farmers. The number of animals kept by a household depends closely on the quantity of fodder that is possible to generate. In order to assist the upkeep of their herds, the farmers try to exploit all sorts of fodder resources, whereby all kinds of greenery constitute potential fodder. They take great care to see that no fodder is left unused or wasted.

An important source of winter fodder derives from the already mentioned *ol* and the grassy banks between the fields or along irrigation channels.[12] Crop

---

12 Wild grass or mountain grass is not cut in Shigar as it is practised in some parts of Tibet and Nepal (cf. STEVENS 1993: 164).

residues are the other important source of fodder, while pure fodder plants are rarely grown in Shigar. In autumn, when the livestock are driven down from the high pastures, they are turned onto the fields to graze on the stubble. Dried apricots of lower quality are given to them in the evening as a kind of reward so that they return to the stables. What kind of fodder is fed to the livestock during the winter depends not only on its availability but also on many beliefs about the appropriate forms and amounts of fodder to be given to particular types of animal. Cattle and crossbreeds are fed straw, and sometimes a little grain. Pregnant cattle are given supplementary fodder, hay, some flour, apricot nuts, and sometimes eggs. Generally, all the livestock lose weight during the winter. Since the traction power of *zo* is needed for ploughing they are fed barley and sometimes even apricot oil as supplement for the last three weeks before ploughing starts in early spring. Goats and sheep are fed leavage of fruit and other trees during October and November. Women or children tear leaves off the trees or beat the branches with a long stick so that the leaves fall down and can be eaten by the stock. During the winter, goats are fed hay, dried leaves and herbs from weeding, sometimes apricot kernels of low quality and also millet straw, when available.

*Communal Herding Management*

As already stated, all sheep and non-lactating goats from a village in Shigar Proper are moved to the high pastures at a fixed day by some villagers while cattle are driven to the high pastures individually by the stockholders themselves. On the *bloq*, however, the livestock is tended by a fixed number of villagers who work there as herdsmen on a rotational basis (*ress* = turn-taking). In contrast to other regions of the Karakorum and even most parts of Baltistan nobody from Shigar Proper stays on the *bloq* for the whole season. Generally, there are two types of pastures or pastoral settlements. Firstly, those for sheep and non-lactating goats, secondly, those for cattle and lactating goats. The rotation system for cattle and lactating goats is called *baress* with a term of twelve days, while the turn-taking period for sheep and non-lactating goats (*luress*) varies between five and ten days.[13] In other villages of Shigar the turn-taking period is based on the number of livestock owned by different households. Farmers look after the whole village's livestock and dairy production for a period corresponding with their household's number of animals relating to the total number of livestock tended on the pasture (GLOEKLER 1995: 16).

For Shigar Proper the *ress*-system should be illustrated by the example of Zembloq affiliated to the villages Agepa, Astana, and Shopa (cf. Fig. 2): In summer 1997 two men were looking after 130 sheep and 35 non-lactating goats at Kachaharal (3,670 m; *haral* = stable, fold). In nearby Blaqbuharal (3,800 m) six men were engaged in tending and milking cattle (3 *hyaq*, 80 *zo*, 80 non-lactating

---

13 The turn-taking for moving sheep and goats in winter to the winter pastures is also called *luress* with a rotation period of two to three days.

*zomo*, 7 lactating *zomo*, 40 lactating cows, 100 non-lactating cows, 30 calves, 4 bulls) and 83 lactating goats (Tab. 3).[14] Each man was responsible for the daily milking of 20 goats, one *zomo* and a few cows.

Table 3: Examples of pasture settlement herds and flocks of Shigar Proper

| High Pasture (*Bloq*) | Connected Villages | Herds-men | Zo | Zomo | Bulls | Calves | Cows | Goats | Sheep |
|---|---|---|---|---|---|---|---|---|---|
| Zembloq-Blaqbuharal | Agepa Astana Shopa | 6 | 80 | 80 | 4 | 30 | 100 | 80 | – |
| Zembloq-Kachaharal | Agepa Astana Shopa | 2 | | | | | | 35 | 130 |
| Thaserpho Tapsa | Ghzoapa | 10 | 100 | 70 | 5 | 50 | 120 | 220 | |
| Abas | Ghzoapa | 4 | | | | | | 80 | 320 |

Source: Own survey, 1997–98.

## Share-Tending Arrangements

The herding management on a rotational basis implies that all households sending their stock to the summer pastures have to tend livestock and manufacture dairy products there for a specific number of days to achieve the right to receive their full share of dairy products (butter) according to the number of their lactating stock on the *bloq*. Several households, however, give their livestock into the custody of friends, neighbours or relatives who participate in the rotational herding system. At the end of the season, these share-tenders are given a service charge. In case of non-lactating animals, the share-tenders are given tea, sugar, money or other goods from the owner of the livestock as payment. For lactating animals the periods as well as the payment are clearly laid down. In Markunja the "share-milking" period is fixed at four months and ten days, and runs roughly from 10 June to 20 October. Usually, the livestock is driven down at the end of September, but the herdsmen keep them until the four months and ten days are over. Only then they are handed over to the owner. According to an informant of Halpapa the following amounts of butter are taken by the owner at the end of the summer, the balance going to the share-milker:[15]

---

14  Some calves are apparently slaughtered just after birth so that less milk is drunk by them which can then be used for the production of butter.
15  The amount of service charge which is retained by the custodians varies from case to case; in Hashupi, for example, it is about 50 % of butter produced (GLOEKLER 1995: 17).

|  |  |
|---|---|
| *zomo* | 13 – 14 kg |
| cow | 5 – 6 kg |
| goat | 1 kg |

If animals are killed due to illness, accidents, predators or other natural events the herdsmen do not have to pay any compensation to the owner. They should just try to use the meat if possible and should inform the owner so that he can take it down to the villages. If animals have to be slaughtered due to injury or illness the herdsmen are allowed to take heart and liver while the rest of the meat is given to the owner. Animals born in high pastures belong to the owner of the dam; he usually gives a small reward to the herdsmen. All animals are marked by scratches on their horns, paint spots on the hide, marks in the ear, or by cloth ear-tags to identify them. This herding arrangement on pay basis is practised when a particular household is short of men or does not want to send a household member to tend the livestock on the *bloq*.

### Communal Breeding Arrangements

The major reason for keeping *hyaq* in Shigar can be seen in their reproductive capacity. There is usually only one grown *hyaq* in each village of Shigar Proper that is used for breeding purposes. It is owned by the whole village community and usually bought at the age of two to three years in the Thale valley (cf. JENSEN 1997: 51). Other villages of the lower Shigar valley buy *hyaq* in the Bashe and Braldo valleys. Only *hyaq* at ages between four and eight years are suitable for cross-breeding while old ones are sold, and from the profits a new younger *hyaq* is bought. During winter such a "village *hyaq*" is held within the stable of one household which receives one basket (*chura*) of straw from every household for the feeding of the animal. In former times, when a village *hyaq* was used to cover a cow or *zomo* the holder of the latter animals had to pay twelve eggs to the *hyaq* holder.[16] Today, farmers from the same village do not have to pay anything for siring their own animals by the *hyaq* while people from other villages without an own village *hyaq* must give some money for that purpose.[17] A different situation occurs in summer when the cattle are on the high pastures. There are specific pastures (e.g., Gongser, Kyaltoro) where *hyaq* and non-lactating cows graze together more or less unattended. The crossing of the animals is thus made easier and is desired. The coverage of cows by a local bull is free of cost. Just in case the village community has bought a breeding bull of a specific race, e.g., by development agencies, people have to pay some money to the community.

---

16   The same method with a village *hyaq* is practiced in Zangla (Ladakh) too, where the holder gets 1 kg butter as payment for each crossing (FRIEDL 1983b: 244).
17   Hashupi villages do not possess a *hyaq* of their own. Due to the costs of siring their cows by a *hyaq* of nearby villages they "prefer to backbreed their *zomo* with common bulls." The resultant offspring *tol / tolmo* is kept in Hashupi in great numbers (GLOEKLER 1995: 15).

## Herding Patterns on the Bloq

The different pasture settlements may vary significantly in size and construction but all of them show the similar features (Fig. 4). Such a permanent pasture settlement consists of one or more clusters of small stone-built living huts (*kuru*) with associated cattle-sheds and pens (*haral*) for the livestock. They are situated just below the main grazing grounds, mostly with a permanent stream close by. The huts are relatively small, irregular in shape, windowless, and with a low roof pole, rafters and stone slabs covered with turf. Some *kuru* are one-roomed, others have two rooms, one for sleeping, cooking, and eating (*mephus*), the other for milk processing (*bjonkuru*). They are sparsely furnished with the minimum needed – blankets, a few cooking pots, dairy utensils such as milking buckets, aluminium pots, butter churn, plastic bags, and the required food for the stay (flour, tea, sugar) that is brought up by each person himself. Small, closed cattle-sheds for calves (*bukuru*) are grouped around the main hut, as are half-open pens for goats (*rakhor-haral*) or sheep (cf. Photo 2). Unlike other regions in the Karakorum (cf. FRIEDL 1983a: 73; OSMASTON 1994: 239), stables and huts are not privately but commonly owned and are used by the whole village community.

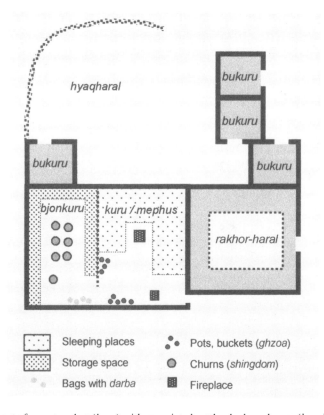

Fig. 4: Layout of a pasture hut (*kuru*) with associated cattle-sheds and pens (*haral*) in Shigar

Photo 2: Pasture settlement near Sesko Village (Bashe Valley) (26. 5. 97)

Photo 3: Milking in early morning at high pasture of Shigar (16. 8. 98)

In early morning, all animals are milked by hand into a small wooden bucket (*ghzwa*)(Photo 3). Goats are milked within the *haral*, cows and *zomo* outside near the stables. This work takes between thirty minutes and an hour. Goats are counted every morning when let out of the *haral*. One or two men, depending on the herdsize, move the goats to higher grounds, while the other men start processing the milk (*oma*) from which butter (*mar, bjon*), the main product of the pasture economy, is made. Although *hyaqmo* produce milk with the highest butterfat content they cannot be milked when on the high pastures where they graze on high grounds. Therefore, the milk from *zomo* which is of similar high quality is much preferred. Greater milk output clearly favours the *zomo* against the *hyaqmo* and common cow. A daily yield in summer is about 2–4 litres of milk from *zomo*, 1–2 litres from the common cow, and 0.2–0.4 litres from goats.[18] Raw milk is usually not drunk, except in salt or sweet tea.

All milk from *zomo*, cows and goats is mixed and collected in pots. It is heated at once in an aluminium pot, a little buttermilk is added and then it is stored for some hours, by which time it will have curdled into a kind of yoghurt. It is then churned with the help of a wooden paddle (*skorbu*) and churn (*shingdom*) or, for smaller quantities, with a leather bag (sheep or goat hide) (*chhanghrkyal*) that is shaken on the knees to make butter. The shaft of the wooden paddle stands vertically in the open-topped churn, close to a roof-pole of the hut, turning easily in a loop of bent willow twigs. This is then turned by pulling in alternate directions with each hand on a wooden toggle wound round the ends of a leatherband a few times. After the churning, the butter is put into cold water to harden it and is then collected in old tin boxes or leather bags. The churning takes around an hour, and different men churn during the morning until noon, by which time this activity is completed. Butter is then stored by the owner in an individual compartment. After the butter has been carried down to the villages it is then cooked again to clean and concentrate it. Butter constitutes an important part of the diet. It is perhaps most often consumed in tea, for it is an essential ingredient of the basic beverage of the region, the *namkin cha* or salt tea that is consumed in great quantities daily by all households. A smaller part of the butter is used for cooking purposes. The making of butter is also an important strategy for storing milk because it can be kept nearly all year round in Baltistan's cool climate. Today, butter is only rarely stored in the ground for years, which is offered to guests as a delicacy. Most of it is kept in little wooden pots (*kase*) and consumed within a year. According to local informants, *zomo* produce about 20 kg of butter annually per head.[19] Usually, the butter is produced locally from the milk

---

18  In addition to the great difference in fodder between summer and winter months, aspects of the animals such as age, suckling, and year of lactation result in a great variation in milk yields. Furthermore, estimations are made difficult by the uncertainty of how much is consumed by suckling calves. FRIEDL (1983a: 68) estimates the lactation yields of *zomo* as 300–400 litres per year; OSMASTON et al. (1994: 219) gives similar numbers with 300–450 litres annually per *zomo*.

19  JENSEN (1997:60) came to similar results.

obtained from the domestic animals, but most households of Shigar Proper need to buy additional butter or fat.

After the butter is separated, the buttermilk is reheated to boiling point which gives a kind of curd (*darba*). *Darba* is put into bags (nowadays in the form of permeable plastic flour bags) which are hung above the ground to drain. Some of the curd is eaten fresh by the herdsmen themselves, but most of it is taken down to the valley when the men are replaced at the end of a rotation circle. Since *darba* is perishable, some curd is given to neighbours, friends and relatives for direct consumption. However, there is no strict rule for the distribution; the owner of lactating animals has no fixed right in receiving any *darba*.[20] During the winter, the stock produce far less milk which is just enough for the direct consumption in tea. Butter and *darba* are thus not produced in winter. In spite of the fact the people in Shigar domesticate so many animals, the returns from pastoralism in the form of dairy produce are rather small. This is made more attractive, however, by the high esteem and sale price of the butter.

*Zo*, non-lactating *zomo* and calves are usually not tended by herdsmen and graze on different meadows. They are driven to the high pastures in late spring and are caught in autumn to bring them down to the villages. *Hyaq* and *hyaqmo* also graze untended on high ground (4,300–4,600 m); they stay in the highest pastures close to or beyond the snowline. Some of them cross into neighbouring valleys where they have to be caught and brought back. The same holds true for *zo* that are sometimes difficult to find since they frequently change pastures. It is not uncommon that men have to search for their cattle over several days until they catch them.

Cows and *zomo* graze on the meadows above the pasture settlements. They return to the pasture settlements every evening for milking and stay the night unattended near the stables, or fenced in an open *haral*. Goats are penned in *rakhor haral*, which is always half roofed, so that the animals have shelter when the weather is bad. In the evening, all lactating animals are milked again. The milk is stored in pots and processed only the next morning. Butter is usually carried down to the valley only at the end of the herding season in autumn.

As opposed to other places such as Askole in Braldo (MacDonald 1994) where men and women go on high pastures or to Hushe where only women work on the high grounds (Emerson 1986: 94), the summer pasturing in Shigar Proper is an exclusive men's province.

*Customs and Traditions*

In former times, several customs were practised in connection with pastoralism in Shigar. Today they are almost never practised or are hidden since these traditions have their origin in pre-Islamic times and therefore are not appreciated by

---

20  In contrast to Yasin (Herbers & Stöber 1995: 100) and Zangskar (Osmaston et al. 1994: 217) the curd is not dried in Shigar, but only consumed fresh.

religious leaders. Today it is difficult to achieve any information about such customs. The following aspects are given by two old men from Gianpa (Shigar Proper).

During the night before goats and sheep were moved to the high pastures people burned juniper branches and walked anti-clockwise around their stables three times. In the early morning when the animals were let out of the stables the head of the household (*nang-go*) stood with burning juniper branches and birch bark at the right hand side of the gate so that the animals had to cross the fragrant smoke of these small fires. It is said that juniper is a sacred plant and its smoke has a cleansing effect on the animals.[21] In addition, in wealthier households, children stood at the gate with plates full of flour. The flour then was given to the poor.[22]

In July all sheep were moved down to the valley for shearing. Men, women and children went with musicians up to Kaniskar, where they received the sheep driven down to there by herdsmen. The music played the whole time as the musicians followed by the flock and the villagers went to the villages. The sheep were sheared over the course of two or three days in the villages and were then driven to the high pastures again. Flour was given to the poor of the village. Today, sheep are sheared only once a year in spring or autumn.

After the harvest of buckwheat in October, when sheep and goats were driven down from the *bloq*, the whole village population and musicians went to Kaniskar again to welcome the animals. All moved to the villages where every household at the central square (*changra*) had to pick out its animals that were marked before sending them to the *bloq*. Today, almost none of these customs remain.

## 5. PASTURAGE RIGHTS

Two land settlements were carried out in Baltistan in 1902 and 1914 by the Dogra in order to have a foundation for proper taxation according to the size and quality of fields. This has led to a distinction between settled land (fields, villages and their surroundings) and unsettled areas (barren land between villages, mountain areas, glaciers etc.). The latter, including the pasture areas, has been declared as government land. However, there are old traditions of village territories and connected grazing lands which continue to guide herders today. Each village or village *mohalla* (section) in Shigar has its own affiliated communal pasture area.

---

21 The special role of junipers as a sacred plant in connection with pastoralism is documented for the whole Himalayan range; from the Kalasha of the Hindukush (PARKES 1987: 649ff.), the people of Bagrot in the Karakorum (SNOY 1975: 112; 1993: 57), to the Sherpas of the Khumbu Himalaya (STEVENS 1993: 176).

22 A similar tradition is practised up to this day: When family members leave the house for a trip to the town or elsewhere, while crossing the gate they touch the flour of the plate with their hand and tip the fingers on their forehead. The flour is also given to poor households of the village.

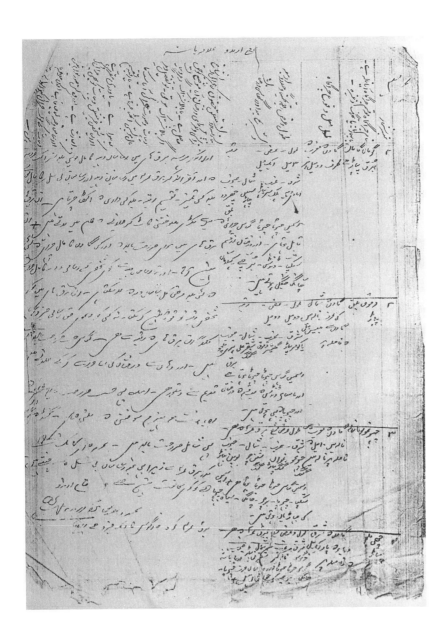

Photo 4: Page of a pasture rights document (*Naql-kajera-i*)

Pastures are considered commons and customary pasturage rights are shared by one or several village communities. In general, the pastures are distinguished between *maqbloq*, which are pastures on which several villages have grazing rights, and *nangbloq*, pastures used by only one village community. In the case of Shigar Proper, Rgyeltsa near Thale La is accessible for all village communities, whereas pastures such as Thaserpho Tapsa or Zembloq are affiliated to a particular village community that has exclusive pasture rights (cf. Fig. 2).

The pasture rights (*Naql-kajera-i*) of the different villages are written down in the revenue records that are used up to the present day as important documents in lawsuits. They were first prepared in a written form under the Dogra during the settlement and are based on traditional customary rights and practices. A few of these documents, by now in a very bad condition and almost unreadable, were able to be copied and translated (Photo 4). The papers inform about the following aspects:

Name, location (in distance from the respective village) and size of all pastures affiliated to the particular village are mentioned. The boundaries of the grasslands are listed exactly in addition to the kind of grass and trees grown there. Statements from villagers about grazing, wood cutting and wood collecting practices are given in another paragraph. For example, it is mentioned that the grassland is good for grazing the animals but too small for being cut and collected, and that the land is unsuitable for cultivation. Furthermore, this paragraph informs about an eventual collaboration between villages in using the same pasture, and states whether a rotation-system is used, or whether farmers from other villages pay any compensation to the village to which the pasture is affiliated. It is clearly stated that nobody from other villages is allowed to graze his animals on the pasture without the permission of the villagers, and nobody can claim for ownership in these pasture areas.[23] Another paragraph deals with disputes and regulations. The dispute is explained or if there is none, that is stated too. Subsequent to these paragraphs the Naib Tahsildar (low-level revenue officer) attests and comments the report and statements given by the villagers. Usually he just recapitulates the statements, and suggests enabling the villagers to use the pastures as stated above. At the end of the document the Tahsildar (mid-level revenue officer) attests the report, his and the villagers´ signatures certify the report.

As assumed and also noted in the reports, the written pasturage rights follow the traditional usage rights practised in the area for generations. The historically grown systems of resource allocation guaranteed that accessibility to grazing lands are equitable for each village within an oasis. However, disputes between village communities about pastures and grazing rights are not uncommon. Espe-

---

23 It is said that before the settlement, some fields and meadows on several high pastures of Shigar Proper were privately owned by some *qaum* (patrilinear lineage). Thaserpho Tapsa for instance, was the private property of the Ghzoapa Wazir. Furthermore, the Bontopa Wazir owned land in Karapur, while some farmers of different villages had fields in Bangma Haral. With the settlement, these areas became government land and all villagers have from then on the right to use these lands as pastures.

cially near the main rivers where a change of riverbeds frequently open possibilities for new pastures, the situation of usage rights is not clearly laid down. As noted above, winter pastures are meagre and limited. As far as Shigar Proper is concerned, only Marapi and Kothang villages have affiliated winter pastures – the villagers of Marapi use the common lands at the other side of the Shigar River, while the livestock of Kothang graze near Shigar River and Jerba Tso. All other villages of Shigar Proper have no affiliated winter pastures. It is worth mentioning here that the grazing rights for summer and other pastures are not identical with the administrative subdivision of Shigar Proper into different *mouza*. In the case of Ranga and Risarpi Haral, for example, villages from Markunja (Halpapa, Daskor, Senkhor), Stotkhor Kalan (Kiahong), and Thugmo have grazing rights to the same pasture.

The trails (*balam*, *bangstrangbu*) leading to the high pastures require building and maintenance, an undertaking which – like the maintenance of the irrigation system – calls for large coordinated labour inputs. Again, the *tsharma* of the respective villages are responsible for calling their villagers to such communal work (cf. POLZER & SCHMIDT 1999).

## *Privately Owned Pastures*

The situation for the mentioned transitional pastures in the Shigar Nalla (Kaniskar, Yulba, Lhchaxmachan) is somehow different. These grasslands are irrigated meadows and included in the settled areas. The ownership is vested in the name of an individual. People who tend their animals there do this according to traditions or to an agreement with the landowner. As an example, the meadows at Yulba are owned by households from different villages. Farmers from Ghzoapa use these meadows as transitional pasture in spring and autumn. They used to give the landowners one basket (*churong*) of manure as payment and helped them by cutting the grass. As said by one owner, today the farmers neither help cutting the grass nor give any manure to the owners. Since all the meadows in this area are not fenced, the owners cannot exclude their own land from this practice and probably do not want to. The grazing of the animals improves the fertility of the land and is thus in the interest of the landholder, too. Another arrangement between the Wazir Ghzoapa and villagers of Agepa, Astana and Shopa about a large meadow in Lhchaxmachan (2,650 m) owned by the Wazir is still in use. The farmers tend their livestock on the grassland in spring and autumn and cut the grass in September which is given to the Wazir as compensation.

## *Land Revenues and Consequences on the Bloq*

In former times, in addition to their own needs, the people of Shigar had to keep more lactating animals to supply butter, milk, wool and manure to the ruler at that time. It was obligatory and was done for the raja as well as for the Dogra officials in connection with the *ress*-system (cf. MACDONALD 1998b).

Furthermore, some of the high pasture areas of Shigar Proper were used for the growing of barley. Not only the suffix -*tapsa*, meaning growing of grain, for the pasture names Thaserpho-Tapsa or Bizerpi-Tapsa give proof of this fact but also the still existing terraces near some alpine settlements. According to local information the growing of barley at these places was done to achieve an additional food resource since the rulers, raja or Dogra, demanded a high tax burden. It is said that in the past, whole families went over several months to the high-elevated areas. This can be seen as advantageous for growing grain, which would be difficult within a rotational system. It is not known exactly in which year the cultivation of barley on the high pastures was given up. Locals report it happened in the 1920's – perhaps with the introduction of the rotational system.

## 6. PRESENT AGRO-PASTORAL CHANGES

During the last centuries the combined mountain agriculture guaranteed the subsistence of the valley's population. The dramatic population growth within the last decades (cf. POLZER & SCHMIDT 1999) in connection with limited growth of productivity and production in order to broaden the food base have led to the necessity of off-farm incomes to cover purchases of much needed goods for the livelihood of the resident population. Means of achieving non-farming incomes are similar to other parts of the region (cf. KREUTZMANN 1989; EHLERS 1995): labour migration, employment in the army, services for the government, or in the tourism sector. Numerous mountaineering expeditions and, to an increasing degree, trekking tours have been carried out annually in the region over the last decades. Since Shigar can be described as the main gateway on the route to the Baltoro Glacier – the destination of most tours and expeditions[24] – these undertakings offer the local population opportunities to work as porters or trekking guides. Households over the entire valley send more or less regularly one or more male members to work as porters on these tours.[25] The consequences are manyfold and include not only the generated cash income, but also the absence of male labour from the villages during the labour-intensive summer months.[26] Two possibilities are conceivable for the affected households to be able to cope with the labour shortage: Firstly, other family members take over the work resulting in a higher workload for each individual. Secondly, the expenditure of labour is reduced, e.g., non-growing of a second crop, decreasing herd size, or renting out of fields or livestock. Regarding the first aspect, large households or extended families can usually compensate the absent working force; brothers fulfil the

---

24  $K^2$ (8,611 m), Broad Peak (8,047 m), Gasherbrum I (8,068 m) and Gasherbrum II (8,035 m) are located at the upper end of the Baltoro Glacier.
25  About expeditions and portering in Baltistan, cf. GRUBER (1981) and MACDONALD (1998b).
26  On the other hand, porters are needed mostly in May and June, when the expeditions move to their basecamps. Less porters are needed during harvest in July and August when the workload in the agriculture is highest, as only trekking tours are carried out during this time and less porters are necessary for the return of expedition groups.

tasks of the man working as a porter. In smaller households, the absence of a male family member often creates serious problems for the remaining people. Women frequently have to replace the absent labourer and carry out work that is usually carried out by males. Consequences of this are not only a higher workload but also a bad reputation for the specific household. However, to conclude that the workload for women increases in general when men work elsewhere as it is often the case, is somehow questionable in the case of Shigar Proper. There is a dualism between the actual higher workload for the remaining family members and the increasing influence of the religious leaders who strictly define which kind of work can be carried out by women and which not. As examples, the high pastures became an area exclusively for men, and irrigation formerly done by both men and women became a predominantly male task. Compared with other places in Baltistan many working tasks are done in Shigar Proper by men that are carried out by women elsewhere (cf. STREEFLAND et al. 1995: 144–145; MACDONALD 1994). The gender-based division of work related to pastoralism is given in Tab. 4.

Table 4: Gender-based division of labour related to pastoralism in Shigar-Proper

| Exclusively or predominantly male | Male or female | Exclusively or predominantly female |
|---|---|---|
| tending livestock and manufacture of dairy products at high pasture | milking livestock and manufacture of dairy products in village | feeding livestock |
| goats shearing, spinning goat hair, weaving and rope making | cutting and collecting grass and fodder | beating and spinning sheep wool |
| butchering, conducting sacrifices | tending livestock in village | tending poultry |
| guarding fields from livestock and wildlife | | weeding fields |
| herding livestock at village pastures in winter | | |

Source: Own survey, 1997–98.

Since work on the high pastures in Shigar Proper is clearly defined as male work, the absence of male labourers results in a setup that small households frequently cannot afford a man to carry out his duty within the rotational cycle. Consequently, such households either give their animals to others for share-tending resulting in a lower butter yield, or they decrease their herd size.

The aforementioned expedition and trekking industry has led to another effect in the villages of upper Braldo: Goats and sheep are sold in high numbers to these groups for a relatively high price (cf. MACDONALD 1994). However, this lucrative additional cash income is restricted to the Braldo valley and therefore has almost no effect on the lower Shigar valley. Much more relevant to the situation in Shigar Proper is the import of meat and animals. Cattle are brought from the Punjab to Skardu and meat and poultry is sold in the local market.

*Prospects of Pastoralism in Shigar*

According to several statements by locals and the above-mentioned findings, the number of animals kept by a single household is definitely in decline. However, due to population growth and a significant increase in household numbers the total number of livestock kept in Shigar is probably not much lower than it has been in the past.

This leads to the question about the role of pastoralism today. Obviously, the necessity to keep livestock for subsistence is decreasing: cattle today are not so much in need for their traction power since only some households still use *zo* and bulls for ploughing, and all threshing in lower Shigar valley is done today by threshing machines (Photo 5). (In Bashe and Braldo, people still use cattle for threshing.) Similarly, goods and people are today transported by jeeps and tractors – most villages are connected by a jeep-track – which means that donkeys or horses increasingly become redundant as pack animals. The dung from animals is still highly valued but almost all households in the lower Shigar valley use

Photo 5: Threshing in Shigar Proper (28. 8. 98)

additional chemical fertilizer that can be purchased in Skardu and even Shigar Proper. Cotton clothes are common nowadays. Clothes made of sheep wool are rarely seen in the valley today. Furthermore, carpets are often bought abroad and replace the local ones made from goat hair or sheep wool. Today, households usually use cooking oil from lowland Pakistan. Local butter is consumed only in salt tea or for traditional dishes that are rarely prepared today. Besides, butter from Bashe is still purchasable in Shigar bazaar but is much more expensive than imported fat.

On the other hand, it is interesting to note that most households of Shigar still keep animals. At least one goat or cow is found in almost all households. Animals are thus mostly kept for their milk supply or to have something in reserve in case of emergencies.

Further conclusions can be stated when looking more closely at the average herd size per household of two villages of Shigar Proper (Tab. 5). There is a significant difference between Halpapa and Daskor when regarding the number of stock per household. While cattle and goats are raised in small numbers in the former, there are much more animals kept in the latter. This fact can hardly be explained by the pasture area since both villages use the same pastures. It is more the result of a changed employment structure in the villages. Approximately 86% of all households in Halpapa have a regular (additional) cash income compared only 48% of households in Daskor. In particular, households with members having a full-time job (not portering) keep small herds. Their fiscal income allows such households to buy the needed products (butter, cooking oil, packed milk etc.) in the bazaar. Furthermore, it is interesting to note that except for one, all households of Halpapa do not keep sheep. No household of this village send male members for herding purposes to the *bloq* today. Thus, Halpapa households with larger herds rent out their livestock in summer to share-tenders, predominantly from Bunpa and Daskor.

Table 5: Herd and flock size per household in Halpapa and Daskor (Shigar Proper)

| Village | Male bovines (*hyaq, zo*, bulls) | Female bovines (*hyaqmo, zomo*, cows) | Goats | Sheep |
|---|---|---|---|---|
| Halpapa | 0.7 | 2.1 | 2.6 | 0.2 |
| Daskor | 1.4 | 3.6 | 6.3 | 3.0 |

Source: Survey carried out by C. Polzer 1994.

A further reason for the changed situation with fewer and fewer people being willing to stay several days on the high pastures is the decreasing attraction of pastoral work. In the past, the availability of plentiful dairy products on the *bloq* and the freedom from the social demands of village life were highly esteemed. Today, butterfat is available all year round in the bazaar, *darba* is regularly brought down to the villages, and several tea stalls, sport, radio and a few TV sets have increased the attractiveness of village life. Furthermore, the prestige of pastoral work is relative low. Wazir families usually do not send family members to the *bloq*. As more and more people try to achieve a higher standard of living and a better reputation – formerly dependent on ownership of land and stock, but nowadays assessed in terms of education, cash income and cash savings –, they avoid taking part in the pastoral work. Consequently, educated persons mostly do not want to work on the high pastures. Finally, the emphasis put on education, particularly for boys who have to attend school regularly, leads to a lack of helpers on the *bloq*. As it can be assumed that in the future less manpower will be available to manage livestock on the pastures due to the mentioned aspects, share-tending arrangements will gradually gain in importance.

## 7. CONCLUSION

Pastoralism has always played a significant role in the overall survival strategies and cultural identity of the people of Shigar. The environmental conditions with limited fodder resources and the close integration of animal husbandry with arable farming confront Shigar's pastoral system with difficult technical and managerial problems. Indigenous knowledge of the range resources that vary considerably with altitude, aspect and season, influence decisions about herd size and composition, as well as seasonal and daily herd movements. Means of resource-management include clearly defined territorial rights of access and utilisation of pastures, communal herding and breeding arrangements. In this way, the Shigar farmers are successful in managing a considerable range of livestock to produce butter, meat and wool, besides the more vital outputs of manure, dung-fuel, and draught power without exploiting its ecological base.

However, the general transformation of the economic structure of the area – a slow shift from subsistence to market economy – leads to an erosion of the traditional form of Shigar's pastoral system. The consequences are manyfold and include the unwillingness of the new social elite to work on the high pastures or the decreasing stock numbers kept by individual households. This does not mean that pastoralism in Shigar is in a definite decline, since it still plays a vital part of local life and economy, but the present tendencies will lead to a redefinition of pasture culture and management styles, including an increase in the importance of share-tending arrangements and a more accentuated differentiation between large and small stockholders.

## 8. REFERENCES

AFRIDI, B. G. (1988): Baltistan in History. Peshawar.
EHLERS, E. (1995): Die Organisation von Raum und Zeit – Bevölkerungswachstum, Ressourcenmanagement und angepaßte Landnutzung im Bagrot / Karakorum. In: Petermanns Geographische Mitteilungen, Vol. 139, No. 2, 105–120.
EMERSON, P. (1986): An Inhabited Wilderness. In: TOBIAS, M. (ed.): Mountain People. Norman, 88–96.

FRIEDL, W. (1983a): Gesellschaft, Wirtschaft und materielle Kultur in Zanskar (Ladakh). (= Beiträge zur Zentralasienforschung, Bd. 4) Sankt Augustin.
FRIEDL, W. (1983b): Landwirtschaft, Viehzucht und Handwerk in Zangla. In: KANTOWSKY, D. & R. SANDER (Eds): Recent Research on Ladakh. (= Schriftenreihe Internationales Asienforum, Bd. 1) München, 237–252.
GLOEKLER, M.-A. (1995): High Pastures – High Hopes. A Socio-economic Assessment of Some High Pastures in Baltistan. Skardu. (Unpublished Report, AKRSP).
GRUBER, G. (1981): Zum Einfluß von Expeditionen und Trekking auf die Umwelt – dargestellt am Beispiel Baltistan / Nordpakistan. In: Wirtschaftliche Aspekte der Raumentwicklung in außereuropäischen Hochgebirgen.In: Frankfurter Wirtschafts- und Sozialgeographische Schriften, Bd. 36, Frankfurt a. M., 21–55.
HARTMANN, H. (1968): Über die Vegetation des Karakorum. 1. Teil: Gesteinsfluren, subalpine Strauchbestände und Steppengesellschaften im Zentral-Karakorum. In: Vegetatio, Vol. 15, 297–387.
HARTMANN, H. (1972): Über die Vegetation des Karakorum. 2. Teil: Rasen- und Strauchgesellschaften im Bereich der alpinen und der höheren subalpinen Stufe des Zentral-Karakorum. – Vegetatio, Vol. 24, 91–157.
HERBERS, H. & G. STÖBER (1995): Bergbäuerliche Viehhaltung in Yasin (Northern Areas, Pakistan): organisatorische und rechtliche Aspekte der Hochweidenutzung. In: Petermanns Geographische Mitteilungen, Vol. 139, No. 2, 87–104.
HEWITT, Farida (1991): Women in the Landscape: A Karakoram Village before "Development". Waterloo. (Unpublished Ph.D. Dissertation).
HURLEY, J. (1961): The People of Baltistan: A Transitional Culture of Central Asia. In: Natural History, Vol. 70, No. 8, pp. 19–27; No. 9, 56–69.
JENSEN, O. (1997): Adaptive Strategies in a Mountain Valley; a Case Study from Baltistan. – Geographica Hafniensia, C8). Copenhagen.
KREUTZMANN, H. (1989): Hunza. Ländliche Entwicklung im Karakorum. (= Abhandlungen – Anthropogeographie, Bd. 44). Berlin.
KREUTZMANN, H. (1996): Yak-Keeping in High Asia. In: Kailash. A Journal of Himalayan Studies, Vol. 18, No. 1–2, 17–38.
MACDONALD, K. I. (1994): The Mediation of Risk: Ecology, Society and Authority in Askole, a Karakoram Mountain Agro-Pastoral Community. Waterloo. (Unpublished Ph.D. Dissertation).
MACDONALD, K. I. (1998a): Rationality, Representation, and the Risk Mediating Characteristics of a Karakoram Mountain Farming System. In: Human Ecology, Vol. 26, No. 2, 287–321.
MACDONALD, K. I. (1998b): Push and Shove: Spatial History and the Construction of a Portering Economy in Northern Pakistan. – Comparative Studies in Society and History, Vol. 40, No. 2, 287–317.
MANN, R. S. (1986): The Ladakhi: A Study in Ethnography and Change. Calcutta.
MIEHE, S., CRAMER, T., JACOBSEN, J.-P. & M. WINIGER (1996): Humidity Conditions in the Western Karakorum as Indicated by Climatic Data and Corresponding Distribution Patterns of the Montane and Alpine Vegetation. In: Erdkunde, Vol. 50, No. 3, 190–204.
OSMASTON, H., FISHER, R., FRAZER, J. & T. WILKINSON (1994): Animal Husbandry in Zangskar. In: CROOK, J. & H. OSMASTON (Eds.): Himalayan Buddhist Villages. Environment, Resources, Society and Religious Life in Zangskar, Ladakh. Bristol, pp. 199–248.
PARKES, P. (1987): Livestock Symbolism and Pastoral Ideology among the Kafirs of the Hindu Kush. In: Man (N.S.), Vol. 22, 637–660.
POLZER, C. & M. SCHMIDT (1999): The Transformation of Political Structure in Shigar Valley/ Baltistan. In: DITTMANN, A. (Ed.): Mountain Societies in Transition. Contributions to the Cultural Geography of the Karakorum. (= Culture Area Karakorum, Scientific Studies, 6). Cologne. (in press).
RIZVI, B. R. (1993): The Balti: A Scheduled Tribe of Jammu and Kashmir. New Delhi.

SNOY, P. (1975): Bagrot. Eine dardische Talschaft im Karakorum. Graz.
SNOY, P. (1993): Alpwirtschaft im Hindukush und Karakorum. In: SCHWEINFURTH, U. (Ed.): Neue Forschungen im Himalaya. (= Erdkundliches Wissen 112). Stuttgart, 49–73.
STREEFLAND, P. H., SHANDANA, H. K. & O. VAN LIESHOUT (1995): A Contextual Study of the Northern Areas and Chitral. Gilgit. (Published by AKRSP)
STEVENS, S. F. (1993): Claiming the High Ground. Sherpas, Subsistence, and Environmental Change in the Highest Himalaya. Berkeley.

# PASTORAL MANAGEMENT STRATEGIES IN TRANSITION: INDICATIONS FROM THE NANGA PARBAT REGION (NW-HIMALAYA)

Jürgen Clemens & Marcus Nüsser

## 1. INTRODUCTION

This paper discusses case studies of agro-pastoral management strategies in the valleys around the Nanga Parbat in northern Pakistan (Fig. 1, 2). These management strategies are embedded in a complex framework of natural pasture resources on the one hand and territorial rights of access and utilization on the other. Apart from the fact, that these aspects can differ spatially, a thorough analysis of the mountain farmers' land use systems together with the potentials and limitations has to integrate the temporal impacts of historical and recent socio-economic processes as well. Indications of external influences and exchange relations, and thus subsequent changes, can be traced back to the middle of the 19th century.

In a first step, such changes can be addressed as being part of the "modernization" process itself that introduced, among others, the improved access and services to previously remote areas. It caused an increased demand for livestock products, the disruption of former trade networks, often an important part of pastoral systems, and the introduction and expansion of national parks and protected areas with further limiting regulations (cf. Miller 1997: 4). These changes occur under a general process of socio-economic and political transition, and the driving forces are located outside the mountains in the Pakistani lowlands ("down country") or on a global scale.[1]

During the 1990s, different studies dealing with processes of socio-economic change and transition in the "Northern Areas" of Pakistan have been published.[2] The aspects and conditions of animal husbandry are mostly incorporated within a broader scope of research.[3] The present analysis of agro-pastoral economies in

---

[1] The authors avoid the term "modernization" because of its implication of particular ideas of development processes or strategies. The analysed changes are discussed under the headline of socio-economic and political "transition", rather than "transformation". For the identification of the dimensions and directions of change in the study area, this concept is more open. This does not, however, regard the area's population as passive objects of a given process.

[2] This list does not attempt to be complete but serves as an overview of important articles: Allan (1986), Kreutzmann (1989, 1991, 1993b, 1995), Khan & Khan (1992), Stöber (1993), Dittrich (1995, 1997, 1998), Pilardeaux (1995, 1997, 1998), Uhlig (1995), Stellrecht (1997).

[3] Kreutzmann (1989, 1993a), Snoy (1993), Clemens & Nüsser (1994, 1997), Ehlers (1995),

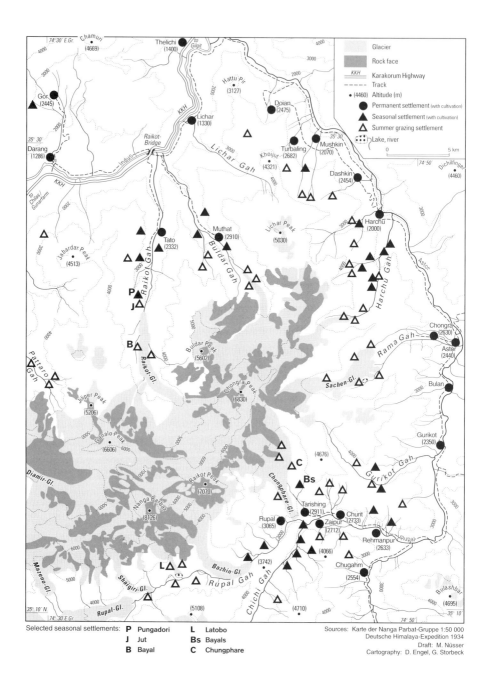

Fig. 1: The Nanga Parbat area

the Nanga Parbat area also includes the mountain farmers' perspective and looks at the dynamics and changes that have occurred, the "agents of change", and at the different impacts on the strategies of resource management. An integrated approach like this also includes changes within the societies and the problem of sustainability of animal husbandry in the different valleys.

For the valleys[4] around the Nanga Parbat Massif, the high rate of population growth and the improved accessibility to the "modern world" via the "Karakoram Highway" (KKH), with the subsequent extension of jeep roads even to remote side valleys, are two major components of the present transitional process. Frequently addressed questions are, whether or not population pressure leads to over-exploitation and subsequent degradation of pastures, whether or not the indigenous strategies of local farmers and pastoralists are flexible enough to cope with new challenges and conditions, and whether or not the new exogenous influences with their new income opportunities lead to definite changes of the "traditional" subsistence economy.

## 2. CONDITIONS OF SUBSISTENCE

### 2.1 Conditions of the Agro-Pastoral Economies

In all mountainous areas of northern Pakistan, animal husbandry is part of an integrated agro-pastoral system which comprises high pasturing, crop cultivation and forest utilization. This "mixed mountain agriculture" (RHOADES & THOMPSON 1975: 537) is characterized by close and interdependent relations between the spheres of livestock rearing and irrigated agriculture. Whereas the availability of manure and animals' drought power is necessary for the replenishment and tillage of fields, the straw and crop residues provide an important fodder after harvesting. Apart from these symbiotic relations within the principal components of land use, the agro-pastoral system requires a complex temporal organization of herd movements in order to utilize the resources of the different altitudinal belts.

The climatic and biogeographical conditions, such as the duration of snow cover on different pastures, the onset of the vegetation period or the dry conditions in the valley bottoms of the colline belt, determine the potentials and limitations of animal husbandry as well as the necessity of pastoral migrations from an ecological point of view. According to the concept of "vertical control" (CASIMIR & RAO 1985: 222), the pastoral mobility extending over the different altitudinal belts assures the mountain farmers' subsistence and allows a wide spread of agrarian risks.

> HERBERS (1998), and NÜSSER (1998a) deal with agro-pastoral economies and particular management strategies in the "Northern Areas". Useful information is also provided by reports by and to the "Aga Khan Rural Support Programme" (eg. AKRSP 1997a, STREEF-LAND, KHAN & v. LIESHOUT 1995).
> 
> 4 The upper side valleys of the research area are locally called *gah* (Shina). The language Shina (Sh.) is commonly spoken by the majority of the population of the Diamir District, including the Nanga Parbat area.

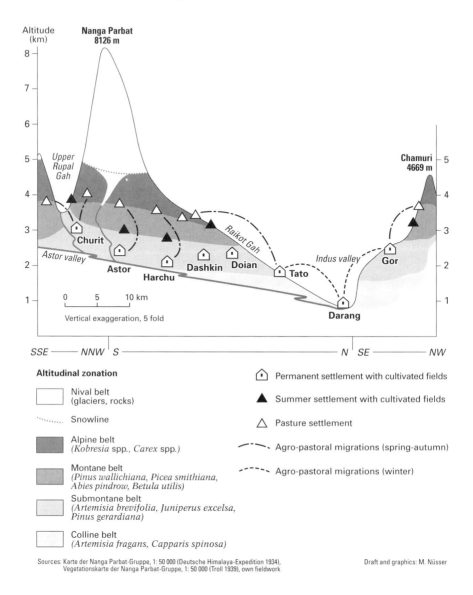

Fig. 2: Longitudinal profile of vegetation belts, land-use and pastoral mobility

Within the high pasture economies, the two fundamental reasons for seasonal migrations to the high pastures of the montane and alpine belts are a lack of pasture resources close to permanent settlements as well as common rules which do not allow the keeping of animals next to the fields during the vegetation period of the cultivated crops. Whereas the upward movement of the herds depends on the snow conditions in the higher belts and on the fodder availability in the cultivated areas, the reverse movement down to the villages can only be deter-

Photo 1: The summer field settlement Bayals (3,200 m) in Rupal Gah (M. Nüsser, 26. 7. 1993)

mined after harvesting. In both cases, the exact dates are fixed by joint decisions of village representatives. These pastoral migrations vary from valley to valley in accordance with the vertical range and the local pasture resources (Fig. 2).

Strategies of animal husbandry around Nanga Parbat are differentiated by three types of agro-pastoral economies. The dominant type is characterized by settled mountain farmers, who usually migrate with their families and herds to high pastures. Generally, the animals are stallfed during the winter with conserved feed supplies, mostly straw and hay, since winter pastures are not available. This type of agro-pastoral economy is described as "Almwirtschaft" (high pasture economy), a term that was originally developed for the European Alps (cf. UHLIG 1995: 201–202 and the introductory chapter to this volume for terminological aspects). To the north of Nanga Parbat, the possibility of all-year-round grazing in the colline belt of the Indus Valley, at least for goats, includes elements of "real" transhumance, according to SCHOLZ (1982: 7, 1991: 31) (cf. also UHLIG 1995: 201). The major characteristics of transhumance – professional shepherds and pastoral movements in winter regardless of the homesteads – are, however, not fully developed here. Additionally, nomadic groups (Bakerwals) from the lowlands of the northern Punjab also reach high pastures to the south of the research area during summer with herds of goats, as well as sheep and donkeys (cf. NÜSSER & CLEMENS 1996a: 129).

Generally, there are two types of seasonal settlements in the upper valley zones: the "summer field settlements" (Photo 1), characterized by supplementary irrigated agriculture, and the "summer pasture settlements", characterized by pastoral utilization only (TROLL 1939: 157–158). These "stages" (cf. UHLIG 1995: 201) are located in different altitudinal zones: whereas the majority of summer field settlements can be found in the montane belt with moist coniferous forests, the summer pasture settlements are mostly to be found in the sub-alpine and alpine belts (Fig. 2). Both types of temporarily inhabited settlements are indigenously called *nirril* or *rung* (Sh.).

Fodder shortages during winter and the frequent lack of forage supplies in spring are the major limiting factors of animal husbandry in the whole region. Therefore, the individual herd size depends on the fodder availability during the cold season. To varying degrees, these fodder shortages occur in all valleys, and different strategies to cope with this main constraint of livestock rearing have been developed. The valleys of Rupal and Raikot may serve as two examples to illustrate the range and the complexity of the stage-wise land use patterns and of resource management in the Nanga Parbat region.

Rupal Gah,[5] to the south of the Nanga Parbat Massif, is a single cropping area with wheat as the dominant crop. Other important crops are barley, cultivated up to 3,340 m, and maize, cultivated up to approximately 2,750 m. Crop residues and wild hay form the major fodder sources for the animals' winter supplies. The principles of the stage-wise utilization of different ecological zones around the year are analysed at the example of Tarishing village (2,910 m) (Fig. 3). Here, the altitudinal distance between the permanent and seasonal settlements reaches approximately 700 m. According to the animals' feed demands and the labour necessities, e.g. of herding and milking, two different pasture areas are mainly utilized. Lactating animals are generally sent to the pastures in the side valley of the Chungphare Glacier in close vicinity of the settlements of Tarishing and Bayals (see Fig. 1). The pastures in the upper zone of the Rupal Gah above Latobo (3,600 m) are mostly grazed by non-lactating animals. For reducing the pressure upon the fodder stores in the village of Tarishing, the *nirril* of Latobo is also used during the cold season. Some shepherds stay there in order to herd hardy sheep and goats all winter long. This is, however, an exceptional case and can only be practiced since strong glacial winds blow off the snow cover on the slopes, and *Artemisia* dwarf shrubs can frequently be grazed (NÜSSER & CLEMENS 1996a: 128, NÜSSER 1998a: 128).

Another adaptive strategy to cope with the frequent scarcity of fodder supplies in winter and spring is to return to the "summer field settlements". Due to limited transportation facilities it is considered easier to guide animals over glaciers (Photo 2) in winter, than to bring fodder supplies down to the villages. Recently, more and more families, however, have stopped this type of winter pasturage. They exchange their fodder supplies of hay and straw with other families between the permanent village and the *nirril*.

5   A detailed description of the environmental aspects of this valley is given by the authors in previous publications (NÜSSER & CLEMENS 1996a, CLEMENS & NÜSSER 1997).

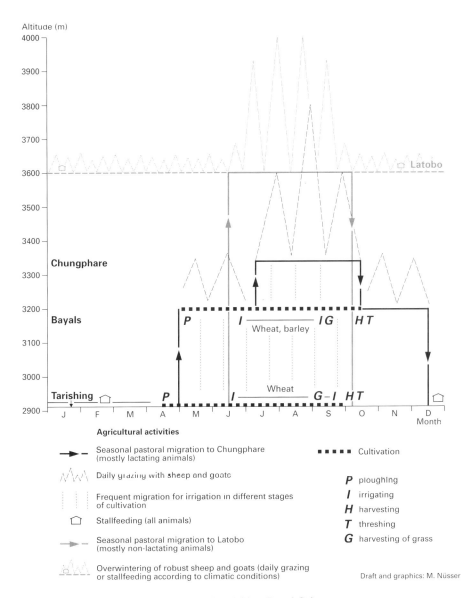

Fig. 3: Stages of land use in the village of Tarishing, Rupal Gah

In the Raikot Gah on the northern declivity of Nanga Parbat the dominant crop is maize, cultivated up to approximately 2,500 m. Additionally, wheat and barley are cultivated up to 3,300 m in the summer field settlements. The majority of households of the Raikot Gah also own and cultivate land in the double cropping areas in the nearby colline belt of the Indus Valley. Double cropping of winter wheat and maize ensures the people's food supplies and additionally higher amounts of crop residues as winter fodder.

Photo 2: Herds crossing debris covered glaciers in Rupal Gah (M. Nüsser, 12. 10. 1992)

The stage-wise land use of high pasturing during summer in the Raikot Gah covers an amplitude of approximately 1,200 m from the village of Tato to the highest grazing settlements. In winter, most of the families descend with their animals to the Indus Valley. Therefore the total pastoral amplitude sums up to more than 2,350 m (Fig. 4). The access to snow free winter pastures in the colline belt of the Indus Valley allows year-round pasturing for the herds of the Raikot Gah and those of the Gor region on the opposite bank of the Indus (NÜSSER 1998a: 139–141). Especially the goat herds are pastured by remunerated shepherds. Stallfeeding is usually confined to cattle and sheep, since the hay production is insufficient for all animals even in the double cropping areas. Thus, this type of agro-pastoral economy may not be adressed as transhumance, but as a modification of the high pasture economy ("Almwirtschaft"), which is dominated by the farming households (see above).

On the slopes around Gor, oak forests (*Quercus baloot*) offer a valuable fodder source in winter. Thanks to traditional rules and regulations, such as the ban on forest pasturing as well as on the loping of branches and leaves for fodder before mid-November, the evergreen oak forests have become an example of sustainable utilization of common property resources.

The temporal development of the forests in the vicinity of a village in the Gor region is demonstrated by repeat photography (Photos 3 & 4). The lower slopes, predominantly exposed to the south-east, are covered with dry forests, dominated

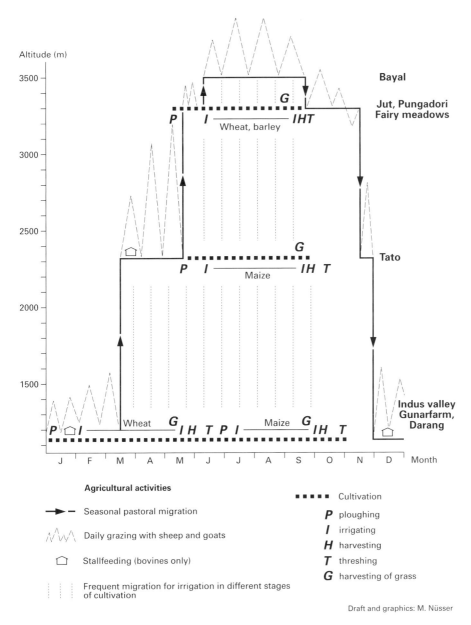

Fig. 4: Stages of land use in the village of Tato, Raikot Gah

by evergreen oaks (*Quercus baloot*) and pines (*Pinus gerardiana*). Apparently, only slight changes in the cover of evergreen oaks in the vicinity of the village can be detected over the period of 58 years.

Photos 3 & 4: Matched photograph pair of the forests of Gor (photographs: above C. Troll, 21.5.1937, below M. Nüsser, 4.9.1995)

## 2.2 Changing Animal Husbandry

With regard to the herdsize, sheep and goats constitute the majority of the animals (Tab. 1), because they offer milk, meat, wool, and skins for domestic consumption and also manure to fertilize the fields. The herdsize itself can also be understood as an emergency asset, and is still orientated towards the subsistence needs of the households. Thus the farmers do not tend to maximize monetary profits and herds are often overaged. KHAN (1991: 10) characterizes this strategy as a "low input – low output-system" and local development programmes aim to improve the productivity of the traditional animal husbandry.

Tab. 1: Comparison of livestock herds per household and composition around Nanga Parbat.

| Species/Year | | Churit 1992 Sources (1) | Tari- shing [a] 1992 (1) | Zaipur 1992 (1) | Reh manpur 1992 (1) | Astor Additional District [a] ca. 1990 (2) | 1992 (3) | 1994 (4) | Raikot ca. 1994 (5) |
|---|---|---|---|---|---|---|---|---|---|
| cattle | nos. | 4.8 | 3.4 | 5.5 | 6.9 | 6 | 3.2 | 5,0 | [a] 10–15 |
| sheep & goats | nos. | 19.2 | 19.6 | 22.7 | 26.9 | 15–20 | 10.4 | 13,0 | [a] 25–40 |
| horses | nos. | 0.2 | 0 | 0.3 | 0.6 | n.a. | n.a. | n.a. | –.– |
| donkeys | nos. | 1.2 | 0.8 | 1.7 | 1.9 | n.a. | 0.8 | n.a. | n.a. |
| horses & donkeys | nos. | 1,4 | 0,8 | 2,0 | 2,5 | n.a. | n.a. | 0,8 | n.a. |
| yaks & zoi/zomo | nos. | | included in 'cattle'! | | | n.a. | 1.1 | 1,2 | –.– |
| others | nos. | n.a. | n.a. | n.a. | n.a. | n.a. | 0.6 | n.a. | n.a. |
| total | nos. | 25,2 | 23.8 | 30.2 | 35.3 | 21–26 | 16.1 | 20,0 | [a] 35–55 |

nos.: number of animals per household; n.a.: data not available; –.–: animals not kept
Sources: (1) CLEMENS & NÜSSER (1994: 384), (2) STREEFLAND, KHAN & V. LIESHOUT (1995: 98–99), (3) BHATTI & TETLAY (1995: 18), (4) MALIK (1996), (5) NÜSSER (1998a: 112).

The herd composition varies with the regional fodder potentials which depend on the variation of the vegetation cover and cropping patterns. Goats dominate at the northern declivity, whereas all over Astor sheep are more important. In the Rupal Gah, the percentage of sheep and goats used to be almost identical, they have changed, however, in favour of goats since the 1970s. This might be explained by the scarcity of winter fodder availability and the goats' resilience. Additionally, the importance of woollen clothes declined. They are substituted by cotton clothes or second hand sweaters and jackets, sold by itinerant traders. Pattu-weaving, the local tradition of wool processing, and recently also carpet and rug weaving are, however, regarded as possible strategies for income generation, e.g. at the regional market places and tourism centres like Gilgit (AKRSP 1997a: 78–79).

The local species of small ruminants are often crossbred with animals sold either by nomadic Bakerwal, especially fat-tailed sheep from Azad Kashmir, or goats, sold by farmers of the Bunar Gah who drive their flocks across the Mazeno Gali (5,399 m) down to the Rupal Gah and to adjoining villages in autumn.

All available data for the Rupal Gah as well as for neighbouring areas indicate that the herdsize per village is increasing. This is due to the population growth, since all families try to keep a minimum number of animals for their subsistence and as an asset.[6] Data for such comparisons, however, has to be analysed carefully. Recent figures from the Astor area reveal that the mean herdsize per household differs significantly between 16 and 26 animals (Tab. 1). The aggregated data of a survey conducted of 15 villages in 1994 also indicate a huge variation within Astor. A more detailed analysis of this data shows that farmers with smaller landholdings keep more animals, based on the ratio of animals to cultivated area, and smaller farmers earn higher proportions of their net income from animal husbandry. Only a low positive correlation was found between herdsize and household size, indicating that households with more people tend to have more animals. The finding can be attributed easily to these households' larger workforce (cf. MALIK 1996).

The recent development trends show significant increases of the number of cattle all over Astor since the partition of Kashmir. Increases in the number of cattle (cows and bulls) and yak-crossbreds since 1947 are reported for the villages in the Rupal Gah (NÜSSER & CLEMENS 1996a: 130, for Yasin cf. HERBERS & STÖBER in this volume). The previous Hindu rulers forbade the consumption of cattle meat also for the Muslim population of Astor (BAMZAI 1987: 44) and pre-Islamic values indicate the neglect of cattle for long periods of time. According to LAWRENCE (1908: 108), the population of Astor did not even use the manure or milk of cows. Nowadays, cattle and yak-crossbreds are preferred because of their higher milk-yields. They also serve as draught animals for ploughing and threshing since topography and the location of the fields do not allow the universal utilization of tractors, especially in higher altitudes and in the summer settlements.

In the course of the summer pasturage, bovines also allow strategies to minimize the workload. Non-lactating animals usually graze free and they are only occasionally provided with some salt, whereas lactating cows and calves are kept and grazed close to the *nirril*. Some villages even divide their animals, leaving the best pastures for lactating cows and calves together with sheep and goats, and non-lactating cattle are sent to higher grazing settlements and pastures (see Fig. 3, cf. NÜSSER & CLEMENS 1996a: 126–127, CLEMENS & NÜSSER 1997: 238, 242–243).

Yak-crossbreds (male *zoi*; female *zomo*) are popular almost all over Astor (Tab. 1), since they are very well suited for the harsh winter conditions and require only limited care during the pasturage season. Yak bulls are mostly kept

---

6   Cf. NÜSSER & CLEMENS (1996a: 130). STREEFLAND, KHAN & V. LIESHOUT (1995: 85, 91) prove this for the entire "Northern Areas", and HERBERS & STÖBER (in this volume) for Yasin. PILARDEAUX's statement (1995: 96–97), concerning the decline of the number of sheep and goats in Astor from 1970/71 to 1993 does not hold true at least for the valleys at higher altitudes, like Rupal.

jointly by several farmers and they can be hired for a fee (*yakluk*) for mating with cows. Due to higher mean-temperatures, *zoi* and *zomo* are not kept at the northern declivity of the Nanga Parbat and the adjoining villages of the Indus Valley. This is an example of different traditional strategies of joint resource management, especially with scarce resources, such as irrigation water, joint herding systems for sheep and goats, or rules and regulations supervised by elected members of the villages with regard to the timing of pastoral migrations or the utilization of common forests.

Concluding it can be stated that on the northern declivity of the Nanga Parbat, the numbers of goats are significantly higher compared with the situation in the Rupal Gah (approx. 25–40 per household against 10–12 in Rupal). Yak-crossbreds are not kept here, and horses are rarely kept due to specific fodder problems. Detailed botanical surveys of the pasture ecology around the Nanga Parbat reveal that until the present day, the intensity of agro-pastoral utilization in the Nanga Parbat region does not show severe degradation of the high pastures (NÜSSER 1998a: 158–163). Regarding the phytomass potential of the natural pastures, even a slight increase in livestock numbers seems to be possible, provided that improved fodder supplies are available in winter.

## 2.3 Territorial Dimensions of the Agro-Pastoral Economy

For a better understanding of the agro-pastoral management systems, it is necessary to include the historical background of utilization rights and recent socio-economic changes, since such complex human-ecological systems with their different interdependencies cannot be fully understood by solely focussing on ecological problems (NÜSSER & CLEMENS 1996a, CLEMENS & NÜSSER 1997, HERBERS & STÖBER in this volume).

The villages in the "settled areas" of the "Gilgit Wazarat", including Astor, were granted rights of grazing and utilization of forests in the beginning of the 20th century. Cadastral surveys ("settlement") were carried out by the *State of Jammu and Kashmir* according to the general practice in British India (cf. LAWRENCE 1875: 400–425, YOUNGHUSBAND 1911: 186–192, SINGH 1917). Historical documents are still kept by the administration in Astor – village-files (*ghas charay*) include the names of different pastures, their boundaries and size, together with specifications of the natural setting. These granted rights of access are still in effect today and determine the villages' actual pastoral economy.

In Astor, the pastures belong to the state and can be utilized as village commons. Mostly, these are adjoining slopes and valley sections of the particular villages. The areas of grazing rights are generally confined by natural boundaries like ridges or drainage lines. Only in a few exceptional cases, villages were granted additional rights in separate areas. The village of Doian in the lower Astor Valley has additional rights of pasturage in the adjoining Indus Valley (Huko Das) and on the slopes north-west of the Nanga Parbat Ridge with the localities of Hattu Pir and Buyar. Some villages also enjoy additional pasturage rights on the opposite river banks. Chongra is another exception. Apart from their rights in the Rama Gah, the people of Chongra also enjoy grazing rights in the upper Harchu Gah, behind the ridge over the Shate Shayn Pass (3,894 m). Different villages in the Rupal Gah share utilization rights in the summer pastures of the upper sections of the Rupal Gah and the Chichi Gah.

The "unsettled areas" along the Indus Valley downstream of the junction with the Astor River show a completely different set of pastoral rights. All valleys north of Nanga Parbat are part to the community of Gor (nowadays Goharabad, Chilas Subdivision), which is located on the right bank of the Indus. All villages of Gor share combined rights of pasturage in the entire area. The people limit, however, their pasturage and pastoral migrations to the slopes and valley sections in the proximity of their particular settlements.

Thus, the main ridge of the Nanga Parbat Massif, from the Mazeno Pass (5,366 m) in the west to the Hattu Pir (3,127 m) in the east, indicates a distinct cultural geographical division, between the village-wise territoriality in Astor and the combined rights of resource utilization in the former, segmentary, "Shinaka republics" of Gor and Chilas. [7]

## 3. INDICATIONS OF SOCIO-ECONOMIC CHANGE AND TRANSITION

The access to the Pakistani communication network via the Indus Valley, especially after the completion of the "Karakoram Highway" (KKH) in 1978, has increased the opportunities of off-farm employment and seasonal migration significantly and has thus exposed the population to dynamic changes.[8] The Rupal Gah and the Raikot Gah have been connected by jeep roads in the late 1980s. Frequent services of jeeps allow young men to work as occasional labourers in Gilgit or Chilas or in the industry and service sectors in "down country" during winter. Simultaneously, the number of tourists has increased with the improved road access, and men take the opportunity to earn money as high altitude porters or as mountain guides for trekking tourists and expeditions.

As part of these exogenous processes, also the modes and intensity of external support changed significantly through internationally funded programmes, facilitating the local population with new opportunities, especially in the sector of "mixed mountain agriculture", resource management, and eventually also in the social sector.

---

7   Cf. General Staff India (1928: 142–143), NÜSSER (1998a: Fig. 21) for a detailed map of the differentiated territoriality. Even today, Nanga Parbat's main ridge forms the administrative boundary between the Chilas Subdivision and the Astor Additional District (until 1993 a subdivision and previously a *tahsil*), both are parts of the Diamir District. In this article, Astor is used synonymously as the administrative unit, the watershed of the Astor River, or the settlement Astor, consisting of administrative offices and the bazaar.
8   The impacts of the KKH are analysed in different studies (cf. KHAN & KHAN 1992, KREUTZMANN 1991, 1993b, 1995, DITTRICH 1995, 1998, STELLRECHT 1997). For major changes of the agro-pastoral systems around Nanga Parbat see Tab. 6 a, b.

## 3.1 Demography and Settlement History

During the last decades, the demographic development in the Nanga Parbat region is characterized by a high population growth, and the total population increased significantly (Tab. 2 & 3). However, already at the beginning of the 20$^{th}$ century, the population pressure in Astor was declared a "Malthusian" problem: "But there is no doubt that the pressure of population is increasing faster than cultivation." (SINGH 1917: 49). The following discussion will present strategies of the local population to deal with this challenge.

Tab. 2: Demographic development in some villages of the Astor Valley.

|  | 1900 | | 1981 | | 1990 | |
|---|---|---|---|---|---|---|
|  | hh. | pop. | hh. | pop. | hh. | pop. |
| Bunji | 89 | 426 | – | 3,012 | 257 | 2,231 |
| Ramghat | 4 | 11 | – | – | – | – |
| Doian | 18 | 149 | 150 | 1,180 | 204 | 1,754 |
| Turbaling | 7 | 49 | 48 | 303 | 47 | 364 |
| Khud Kisht | 10 | 64 | 56 | 399 | 66 | 555 |
| Dashkin | 24 | 229 | 158 | 1,393 | 212 | 1,966 |
| Harchu | 17 | 160 | 140 | 1,109 | 188 | 1,540 |
| Patipura | 21 | 187 | 60 | 403 | 78 | 532 |
| Chongra | 80 | 449 | 333 | 2,201 | 349 | 2,877 |
| Eidgah | 48 | 332 | 216 | 1,580 | 242 | 2,042 |
| Bulan | 11 | 106 | 156 | 502 | 90 | 666 |
| Gurikot | 74 | 538 | 404 | 2,567 | 483 | 3,939 |
| Rehmanpur | 38 | 219 | 191 | 1,360 | 241 | 2,038 |
| Zaipur | 27 | 182 | 94 | 689 | 113 | 933 |
| Churit | 38 | 219 | 191 | 1,360 | 241 | 2,038 |
| Tarishing | 39 | 249 | 191 | 1,387 | 233 | 1,890 |
| Chugahm | 38 | 306 | 263 | 1,988 | 407 | 3,754 |

hh.: no. of households, pop.: no. of population
Sources: Kitab Hukuk-e-Deh (1915/16), SINGH (1917), Government of Pakistan (1984), Revenue Office, Astor (1990).

Tab. 3: Demographic development in selected villages of the Indus Valley.

|  | 1900 | | 1981 | | 1990 |
|---|---|---|---|---|---|
|  | hh. | pop. | hh. | pop. | hh. |
| Thelichi | 6 | – | 32 | 282 | 33 |
| Darang | 6 | – | 55 | 426 | 65 |
| Bargin | 4 | – | 17 | 158 | 29 |
| Gor | 300 | 1,696 | – | – | – |
| Dirkil | 8–12 | – | 129 | 920 | – |
| Gonar | 15–20 | 98 | 350 | 2,350 | – |
| Bunar | – | 640 | – | – | – |
| Tato | – | – | 47 | 345 | 82 |
| Muthat | – | – | 67 | 545 | 73 |
| Lichar | – | – | - | - | 8 |

hh.: no. of households, pop.: no. of population
Sources: General Staff India (1928), Revenue Office, Chilas (1981, 1990).

From evidence in historical records and information derived from the authors' fieldwork, assumptions concerning the settlement history of this area can be made. Along the longitudinal profile of the Astor Valley and the Rupal Gah, different population groups are situated. In the Rupal Gah, the villages of Rehmanpur and Churit are considered to be the oldest settlements. As a reaction to the frequent destruction and desertion of entire villages and the particular cultivated areas until 1851 through the "Chilasi raids", families from Baltistan were invited to settle in Astor (DREW 1875: 399). This has led to the present day prevalence of Balti migrants in the villages of Tarishing and Rupal Pain. As late settlers, these inhabitants of the Rupal Gah occupied free niches in the upper valley sections. Therefore, one can find an overlapping of utilization rights in the longitudinal cross-section of this valley, which is not at all typical for Astor (NÜSSER & CLEMENS 1996a: 122, 126).

In comparison, the villages of the Raikot and Buldar Gah to the north of Nanga Parbat were inhabited from the region of Gor, just at the opposite bank of the Indus, according to the oral traditions. Gujurs, previous nomads who left the lowlands of the Punjab, also settled at the northern declivity and in side valleys of the Indus, especially in the village of Lichar and in different settlements of the Raikot Gah and the Gor area. In the past, the Gujurs were mainly occupied as tenants (*dakane*), and their activities also included the duties of shepherds during the seasonal agro-pastoral migrations.[9] The villages of the Astor Valley have no such records of tenancy. Until the present day, all agricultural work in Astor is performed by family members.

### 3.2 Accessibility and Exchange

Similar to the migration route of the Bakerwals, the present nomadic groups, the resident farmers of the Rupal Gah also used to cross the Shontar Pass (4,564 m) with their own pack animals up to two times per year to supplement the supply gaps of the local subsistence economy. Until the 1970s, they purchased basic commodities in Azad Kashmir or in Mansehra (NÜSSER & CLEMENS 1996a: 123). A similar strategy was also applied by the population of neighbouring valleys, like the Mir Malik Gah (GÖHLEN 1997: 291). This proves two facts: the insufficient agrarian base of the single cropping areas of Astor,[10] and the tradition of

---

9  A British report of 1904 (quoted in DANI 1989: 292) refers to the situation of Gujurs as tenants of the Chilasi population.
10 The same applies already for the Kashmiri period in Astor (SINGH 1917). Nowadays, the "Northern Areas" rely heavily on "imported" wheat supplies from "down country", and the self sufficiency level decreased from eight to ten months per year in the early 1970s to just four to six months in the 1990s (cf. KREUTZMANN 1995: 222, and the introduction to this volume). The governmental "civil supply" started in 1972, subsidizing the transportation of wheat, salt, and sugar (World Bank 1990: 95, DITTRICH 1995, PILARDEAUX 1995). In the village Churit, 103 out of 115 sample households regularly buy wheat or flour from local shopkeepers (survey by J. Clemens, 1992–1994).

exchange with the lowlands contradicting the often proclaimed "closure" or "isolation" of the high-mountain valleys prior to the completion of the KKH (cf. KHAN & KHAN 1992: 13). STELLRECHT (1997: 20) addresses the lowlanders' perspective of the closed or isolated mountain areas as "a familiar cliché". Modern transportation infrastructure, however, intensified the lowland-highland interaction and especially the mobility of the male mountain population and the supplies of food and commodities.[11]

Direct impacts of the improved road infrastructure on the transition of the animal husbandry patterns are hard to find. Nevertheless, a significant indicator might be the decline in the number of horses all over Astor. These pack animals provided an important source of income to the local population after the British and Kashmiri authorities had constructed mule tracks from Srinagar via the Burzil and Kamri Passes down to the junction of Astor and Indus rivers in 1894, the "Gilgit-Bandipur-Road" which was improved in 1925. The households in the Rupal Gah gained more than half of their cash income from renting out pack animals and the authorities provided the farmers with special credits (*taccavi*) to purchase more pack animals. In contrast to the Astor area, horses and mules were not kept in the Indus Valley north of Nanga Parbat (cf. NÜSSER 1998: 82–90). Fodder crops such as lucerne (*rishka* or *ishpeti*, Sh.) also contributed to the cash income of many villages along the "Gilgit-Bandipur-Road" during the British and Kashmiri rule. For the supplies of the governmental transportation, local farmers were also forced to sell hay and straw to governmental depots at fixed rates (*hukmi kharid*) or sometimes as a land tax in kind (SINGH 1917: 110, 126–127, NÜSSER 1998b).

Traffic conditions changed to the disadvantage of Astor after some improvements along the Babusar Pass via Chilas in 1898. This route was further improved by the British after the "Gilgit Lease" in 1935, since it offered a bypass to the territories under Kashmiri control, which also included Astor. With the partition of British India and Kashmir in 1947, Astor has been cut off completely from the trade with Kashmir. Nevertheless, the supplies of the Pakistani troops along the "Cease Fire Line" (nowadays called "Line of Control"), dividing the Pakistani and Indian administered parts of Kashmir, are transported on jeep tracks, built after the late 1950s from the Indus Valley through Astor.

The figures of the official livestock census in 1970/71 indicate the continuing importance of horses and donkeys along the major streams and routes in Astor, especially for the villagers' own commodity supplies (NÜSSER & CLEMENS 1996b: 164, 169–170). Up to the present, donkeys have provided the major source of carriage in the context of agro-pastoral migrations and the farmers' fuelwood supplies, since the high pastures and forests are hardly accessible by jeeps or tractors. Horses, however, are nowadays kept only by few wealthy farmers or as an emergency asset in areas frequently cut off by snow avalanches (cf. HANSEN 1997). The development of the number of horses and donkeys in the

---

11  Food items dominate the trade between the lowlands and the highland, counting for up to 70 percent of its volume (cf. KREUTZMANN 1995: 222, DITTRICH 1997: 28–29).

villages of the Rupal Gah from 1915/16 to 1992 is presented in Tab. 4 (for comparisons with other areas see Tab. 1). In 1915/16, nearly every household in the Rupal Gah possessed one horse. This situation changed significantly in favour of donkeys until the 1990s.[12]

Tab. 4: Development of horse and donkey numbers in the villages of the Rupal Gah.

| Year | Churit | | Tarishing | | Zaipur | | Rehmanpur | |
|---|---|---|---|---|---|---|---|---|
| | *percentage share of equides of the total livestock population (livestock units)* | | | | | | | |
| | Horses | Donkeys | Horses | Donkeys | Horses | Donkeys | Horses | Donkeys |
| 1915/16 | 19.0 | 0.6 | 17.2 | 0 | 20.3 | 0 | 21.3 | 0 |
| 1970/71 | 4.8 | 2.0 | 1.4 | 1.8 | 2.0 | 3.4 | 4.7 | 6.5 |
| 1992 | 0.6 | 4.8 | 0 | 3.4 | 0.8 | 5.5 | 1.6 | 5.2 |

Livestock units: sheep/goats: 0.1; bovines: 0.8; donkeys: 0.5; horses/yaks: 1.0
Data: NÜSSER (1998a: 111), after different sources and fieldwork by J. Clemens & M. Nüsser.

### 3.3 Tourism

Tourism, especially mountaineering and trekking, is a major driving force within the recent transitional process. In the Nanga Parbat area, however, the number of tourists cannot be compared with those in Hunza or Baltistan (GRÖTZBACH 1993: 105). Although there is a tradition of expeditions to the summit and the surrounding peaks of the Nanga Parbat Massif since the 1930s and 1950s, only a limited infrastructure has been developed until recently. It consists of a small hotel in Tarishing (2,911 m) at the road head of the jeep track in the Rupal Gah, fenced camping sites at Fairy Meadows (3,300 m) and adjoining places above the road head in the Raikot Gah, and different sporadic camping sites along the main trekking routes, e.g. near Latobo in the upper Rupal Gah (CLEMENS & NÜSSER 1996: 86, NÜSSER & CLEMENS 1996a: 119). The commercialization of tourism on a larger scale is limited to a luxury hotel by the governmental "Pakistan Tourism Development Corporation" (PTDC) at Rama, above Astor. The "Shangri La" hotel-chain failed to build a mountain resort at Fairy Meadows because of the resistance of the local population (for details of this conflict cf. CLEMENS & NÜSSER 1996). Instead, different families of the village Tato have begun running fenced camping sites from 1992 on.

Due to the better transportation facilities after the completion of the KKH and the subsequent permission for foreign tourists to visit the "Northern Areas", the number of visitors increased also in the Nanga Parbat area. Tab. 5 illustrates this development from 1975 to 1993, concentrating on expeditions that need an

---

[12] Data for 1970/71 does not necessarily reflect the real situation, since this livestock census was conducted prior to the abolishment of land taxes in 1972. Due to the farmers' biased underestimation the real figures are supposed to be higher.

Tab. 5: Expeditions to Nanga Parbat.

| Year | Expeditions total no. | Rupal-Side no. | Diamir-Side. no. | Participants total no. |
|---|---|---|---|---|
| 1975 | 1 | 1 | – | 21 |
| 1976 | 3 | 2 | 1 | 25 |
| 1977 | 1 | – | 1 | 13 |
| 1978 | 3 | 1 | 2 | 14 |
| 1979 | 3 | 1 | 2 | 41 |
| 1980 | 2 | 2 | – | 17 |
| 1981 | 4 | 2 | 2 | 42 |
| 1982 | 4 | 2 | 2 | 47 |
| 1983 | 11 | 4 | 7 | 78 |
| 1984 | 8 | 4 | 4 | 61 |
| 1985 | 12 | 6 | 6 | 103 |
| 1986 | 4 | 1 | 3 | 29 |
| 1987 | 6 | 3 | 3 | 40 |
| 1988 | 8 | 4 | 4 | 61 |
| 1989 | 7 | 4 | 3 | 44 |
| 1990 | 16 | 7 | 9 | 156 |
| 1991 | 6 | 3 | 3 | 41 |
| 1992 | 13 | 6 | 7 | 96 |
| 1993 | 9 | 4 | 5 | 88 |
| Total | 121 | 57 | 64 | 1,017 |

The numbers of porters are not available, the average is estimated between 100 and 150 per expedition and varies with the size, duration, and destination of the expedition.
Source: Alpine Club of Pakistan, Rawalpindi.

official permit.[13] These figures show a slight preference for the northern and western slopes of the massif (the Diamir Side), which holds true especially for individual visitors and guided trekking tours. A trip to Fairy Meadows and the "first basecamp-area" can be done in just two days, because a jeep track links the village of Tato directly with the KKH at Raikot Bridge since 1988.

The implications of tourism with regard to the transitional process of the mountain societies are manifold. The first and still the most widespread one is the hiring of porters and mountain guides. It offers the opportunity of off-farm-employment for men during summer and autumn without the need to migrate. This has led to some modifications of the traditional patterns of agro-pastoral migrations, especially in the villages close to the trekking routes. Due to the job opportunities provided by trekking groups, more young men tend to stay in the permanent villages near the jeep tracks also in summer instead of participating in

---

13 Figures of individual tourists are not recorded properly, especially in hotels and camping sites around the Nanga Parbat. Thus, no comparison with Hunza can be made, where the number of registered foreigners staying overnight increased from 302 in 1979 to more than 5,300 in 1985 (KREUTZMANN 1996: 432).

the upward migration to the summer settlements (NÜSSER & CLEMENS 1996a: 125–126).

This coincides with the practice of an increasing number of families to spend also the summer in the permanent villages, where schools and small bazaars are easily accessible. In these cases, the adult men manage to irrigate the fields of the summer field settlements during occasional visits (see Figs. 3, 4). The animals are then handed over to relatives on the high pastures in return for an agreed share of milk products or, in the case of the Raikot Gah, also to paid shepherds.

Compared with Hunza, parts of Baltistan, and Kaghan (GRÖTZBACH 1993, KREUTZMANN 1996), this recent development has not significantly changed the agro-pastoral economy around Nanga Parbat. Only few marginal pastures are reported to have been abandoned in the Astor region (PILARDEAUX 1995: 97, 103–104). Unlike in Hunza, farmers around Nanga Parbat do not stall feed their lactating animals all year round. The partial shortage of workforce, due to migration and off-farm employment, has not yet led to major changes of the agro-pastoral management and production systems. Still, labour intensive activities continue to be fulfilled within the village communities and in most cases within the individual households. Women are, however, taking over more duties which were traditionally performed by men. In spite of the given exogenous influences, animal husbandry and pastoral migration will continue as key elements of the valleys' economies. High pastures, however, already serve new and external functions, offering recreation for Pakistani and foreign tourists and income opportunities to parts of the local population.

### 3.4 Cash Economy and Labour Migration

One of the major aspects of the recent transitional process is attributed to the money economy and especially to cash income from off-farm employment and labour migration. Cash income from the hiring of pack animals or the sale of fodder was already important in Astor during the last century. Additionally, farmers from Astor were often exempted from the payment of their land taxes in kind. Instead, they were allowed to pay it in cash, because of the frequent crop failures. PILARDEAUX (1995: 51) argues that the tax payment in cash, or grain imports from Kashmir to pay the taxes in kind, proves Astor's early integration in supra-regional exchange patterns.

In Astor labour out-migration already started in the late 19th century, and subsequently, the traditional household economy changed from a subsistence economy to a "remittance" or "mixed economy" (cf. STÖBER 1993, PILARDEAUX 1995: 50 after STALEY 1966, World Bank 1995: 6). Although agriculture and animal husbandry are still held in high esteem among the local population, off-farm employment is the major source of income. Nearly 50 percent of the total net household income in Astor originates from remittances and off-farm employment, although the variation is high.[14]

---

14 The mean data of the percentage of "other", i.e., non-agrarian, income sources in relation to

An overview of different parts of northern Pakistan also reveals the importance of "other household income" sources in relation to net farm income. The data of the mean per capita income for Astor varies between 4,360 Rs. for 1994 and 5,405 Rs. for 1992, which is, however, lower than the mean per capita income in the neighbouring Gilgit District (5,628 Rs.) and of Pakistan (9,170 Rs.).[15] Apart from the presented mean income data, significant wealth differentials within the communities and regional disparities have to be considered as well.[16] Since most reports only provide aggregated data, the percentage of households below the poverty line is generally not addressed (cf. BHATTI & TETLAY 1995, MALIK 1996). According to a separate analysis, the percentage of absolute poor in Astor increased from 39 to 57 percent between 1991 and 1994.[17]

In the village of Churit (Rupal Gah) the majority of households has at least one male member who earns cash income from seasonal or permanent off-farm employment or from pensions. Only three percent of the households reported no off-farm employment. Out of the sample of labour migrants, more than 50 percent regularly migrate to Gilgit or to "down country" for employment or for higher education, which is mostly supplemented with part-time employment. The public sector and especially the army contributes the biggest share of employment, a fact which is verified by other studies as well (cf. KREUTZMANN 1993a: 30, BHATTI & TETLAY 1995, PILARDEAUX 1995: 101, MALIK 1996).

The data on off-farm activities and labour migration proves the population's strategy towards the diversification of their entire subsistence. In central Astor, just 33 percent of the men declared themselves farmers in the 1990-election list (PILARDEAUX 1995: 101), and the percentage of full-time farmers in Hunza is roughly 17 percent (KREUTZMANN 1993a: 30). Since the majority of men in the villages has a "multiple job" employment, supplementing farming with at least occasional labouring, there is also a huge degree of hidden unemployment.

Seasonal labour migration is well adopted to the agro-pastoral land use patterns in single cropping areas. Most of the men only leave their villages in autumn after the completion of the harvest and the collection of fuelwood supplies (NÜSSER & CLEMENS 1996a). Survey results in Churit for the winter season 1992–93 indicate a significant reduction of the mean household-size by 1.2 persons: 68 men out of a sample of 56 households leave their families for two to four months. The situation in the village Batwashi in the Mir Malik Gah is similar: nearly all men spend the winter in Karachi or in other industrial locations (GÖHLEN 1997: 292).

---

the total net income in 16 sample villages in Astor varies between twelve and nearly 75 %. (Data for 1994 based on MALIK 1996).

15 Cf. for Astor: BHATTI & TETLAY (1995) for 1992, MALIK (1996) for 1994, for Gilgit and Pakistan data for 1990–91: World Bank (1995: 20, 22).

16 The village-wise mean data of the net per capita-income in Astor varies between 1,270 and 5,958 Rs. (MALIK 1996).

17 Cf. PARVEZ & JAN (n.d.: Tab. 6); their definition of absolute poor was adjusted to the national one which is focused on the minimum nutritional intakes of the population.

As a consequence, women gain more responsibilities, and they have to deal with a higher work load, especially in small and nucleated families. The seasonal variation of the availability of male labour force leads to some changes in the organization of the traditional agro-pastoral land use patterns. The additional winter visit to a *nirril* cannot be sustained with the same intensity. Smaller households are often forced to send their animals with relatives to the summer pastures, mostly for a payment in kind. Families with permanent employees, e.g. teachers or soldiers, try to combine the men's annual leave with the peaks of agricultural labour. Some wealthier families occasionally hire labourers during the harvest season or buy fuelwood supplies from neighbouring households or villages.

Also due to the increasing school enrolment, especially of boys, the pasturage control is changing. Boy-shepherds are no longer available, and within the joint rotation system of the daily grazing activities,[18] more and more adults and old men carry out this task. A similar process can be observed on the summer settlements, which are visited mainly by women, small children, and older people. Adult men often join their families on the high pastures only for the particular days of their grazing term.

Thus, only bigger households, mostly with three generations, can fully sustain the traditional agro-pastoral activities and also send some of their members to off-farm employment. At the same time these households actively spread their income and subsistence potentials for a high degree of risk insurance. The most vulnerable groups are small households with small landholdings and small herds. Since their subsistence cannot be assured from the traditional mountain agriculture, they are forced to earn cash income, which again limits their agro-pastoral workforce. This often results in a higher share of work that has to be met by the women. One possible adoption to workforce constraints is the change of the herd-composition in favour of cattle, which is less labour intensive (STÖBER 1993: 84, STREEFLAND, KHAN & v. LIESHOUT 1995: 98–99).

### 3.5 Development Activities of Governmental and Non-Governmental Organizations

The impact of governmental policies on the agro-pastoral economy of the "Northern Areas" during the colonial and post-independence periods has already been analysed in this paper. These policies were mostly confined to revenue generation and securing or improving the transportation facilities. A major change occurred after the abolishment of the feudal reign as well as land taxes between 1972 and 1974 under President Z.A. Bhutto, together with the introduction of subsidized commodity supplies to the "Northern Areas".

---

18 STÖBER (1993: 92–94) presents a similar rotational system of joint herding of sheep and goats in Yasin developed after the late 1970s as a reaction to shortages of the boys' and young men's work force. Previously, farmers hired specialized shepherds for grazing.

Except the construction of the major supply routes (primarily with strategic objectives) and link roads, few power stations, schools for boys, and occasional infrastructure schemes in selected villages, no broad-based strategy for the socio-economic uplift of Astor has been developed so far. The application of nation-wide policies or facilities, such as subsidized inputs or agricultural credits, often excludes large groups of Astor's population. This is due to the limited accessibility of settlements or to systematic constraints, e.g. in the banking sector, excluding smallholders from credits (PILARDEAUX 1995), and to management shortcomings. Especially the lacking governmental services, e.g. agriculture, veterinary, or health, are major bottlenecks for the economic development in Astor. Reports and especially the villagers' statements frequently claim the inadequate facilities.[19]

Only after the expansion of the internationally funded "Aga Khan Rural Support Programme" (AKRSP) to Astor in 1993, a broader approach to improve the economic conditions of mountain farmers was introduced. With the general objective of providing market incentives to previous subsistence farmers, this programme focuses especially on self-help activities and capacity building on the village level (cf. KHAN & KHAN 1992, KREUTZMANN 1993b, World Bank 1995, and CLEMENS 1999 for details of AKRSP's activities).

With regard to the livestock sector, AKRSP follows a combined strategy to increase the productivity, including the reduction of animal losses, improvements of the fodder supplies, and of the genetic base of local livestock. Voluntary "village organizations" are offered the participation in different programme packages, after they have agreed to a general agenda.[20]

The lack of veterinary services all over Astor is addressed by the training of semi-professional "livestock specialists"[21] and by the temporary support with vaccines and animal medicines. Farmers are obviously prepared to pay for the specialists' services and the acceptance of this proramme package is higher than in the agriculture or forestry sector. AKRSP also co-operates with the concerned governmental line department to share resources. The governmental policy of free of cost services, however, often causes complications, since the farmers have to pay in cash for the services of the trained specialists.[22] Regular monitoring

---

19 Author's own observations; cf. also AKRSP (1997c) from the villagers' perspective; for the governmental veterinary services cf. World Bank (1995: 51) and FREMEREY, KIEVELITZ & MAHAL (1995: 31).
20 From 1993 to the end of June 1997, 122 "village organizations" and 64 "women's organizations" accepted AKRSP's "terms of partnership" in Astor. However, just 62 or 64 percent of them were recently regarded as "active" (cf. AKRSP 1997b).
21 Until the end of June 1997, 56 "livestock specialists" (male villagers), and 33 "poultry specialists" (female villagers) were trained. Additionally 14 "livestock specialists" attended special courses and serve as supervisors ("livestock master trainers") to the local specialists. However, just about 35 percent of all "livestock specialists" in Astor were regarded as "functional" in 1995 (FREMEREY, KIEVELITZ & MAHAL 1995: 31), against some 70 percent in the entire programme area of AKRSP in 1991 (World Bank 1995: 50).
22 The prices are 5 Rs. per cattle and 2.5 Rs. per sheep or goat, plus the vaccines, against the

reports of AKRSP suggest that the vaccination and medication coverage in Astor improved as a result of the specialists' activities although the mortality performance is still demanding (see below).

AKRSP's approach to improve the genetic potentials of the local herds has undergone a severe change. Artificial insemination is regarded as a complete failure in Astor. Only six out of 82 inseminated cows delivered successfully (FREMERY, KIEVELITZ & MAHAL 1995: 32). Thus, this policy was revised with a new focus on the selection of local breeds instead of exotic ones. Prior, the targets of "improved breed cows" were regularly not achieved (AKRSP 1997b: 21, 26). This new policy is more or less an institutionalized form of the "trial and error" approach that the local farmers have already practised for a long time by buying sheep and goats from neighbouring valleys for crossbreeding with their own stocks.

Based on the farmers' experiences and their frequent demand for improved animal breeds (AKRSP 1997c), there is a potential for quality improvements, provided that the fodder supplies can be improved and marketing opportunities assure returns for the additional inputs (cf. World Bank 1995: 52). There may occur, however, negative side effects like in the Gilgit region, where AKRSP's cattle improvement programme led to a high increase of the women's daily workload (AKRSP 1997a: 49).

In Astor, the livestock activities of AKRSP are concentrated on seven "model villages". An exception is the distribution of alfalfa (lucerne) seeds, which are offered to all "village organizations" (AKRSP 1997b: 7–8, 26). Since AKRSP has until recently focused nearly all its programme activities on the cultivated areas "below the irrigation channel", there are hardly any activities with regard to the improvement or conservation of high pastures. However, approaches towards the control of free grazing and to integrated "natural resource management" are in progress (AKRSP 1997a: 37). Such activities are partially combined with a new programme of "The World Conservation Union" (IUCN). Since 1997 the village communities of Doian, Dashkin, Turbaling and Bunji are directly integrated into the process of drafting and implementing plans on village level for the sustainable use of natural resources and the conservation of biodiversity" (cf. IUCN 1998).

Summarizing, it can be stated that AKRSP is nowadays one of the major "agents of change" all over Astor, at least regarding agriculture and animal husbandry. This programme's funds and facilities outnumber the ones of the governmental line departments and its activities address the farming population directly. AKRSP's activities are, however, still confined to the Astor Additional District. In the remaining subdivisions of the Diamir District, rural development remains with the governmental line departments, and a few local initiatives. As

free of cost supplies by the governmental veterinary dispenseries, which, however, are often running out of stocks and thus are not reliable (FREMEREY, KIEVELITZ & MAHAL 1995: 31–32, World Bank 1995: 50-51). The acceptance of the paid support for the animals' treatment may be explained by the farmers' limited alternatives.

late as the spring of 1998 a similar rural development programme was started in the subdivisions of Chilas and Darel & Tangir by the "International Fund for Agriculture Development" (IFAD) and the Government of Pakistan. This programme covers the "development of agriculture, forestry, and improvements in the standard of education" ('The Dawn', April 2, 1998, Karachi).

Beside the above mentioned policies and approaches, further implications, especially on the "above the channel" elements of the agro-pastoral economies, are to be expected in the framework of new national parks and nature reserves in the Nanga Parbat area. Already in November 1975 parts of Astor were declared "Astor Wildlife Sanctuary" comprising 41,472 hectare. There are no further restrictions for the domestic utilization of forests and pastures, and no management plan existed until recently (GREEN 1993: 404–405). By spring 1995, however, "Fairy Meadows National Park" and "Deosai Plateau Wilderness Park" were established by the "Chief Commissioner" of the "Northern Areas" (MOCK 1995). This contributes to the original proposal of integrating the Astor sanctuary into a much larger national park (GREEN 1993).

Restrictions of access to and utilization of the alpine belts are rejected by the local population who are demanding more participation and "co-management" activities (cf. CLEMENS & NÜSSER 1996), similar to IUCN's activities. This is also based on negative experiences of farmers in the area of the "Khunjerab National Park", who had to fight for their promised compensation at the court for more than 15 years without a final solution (KREUTZMANN 1995: 224-225). MOCK (1995: 19) states that "effective management of the parks of the Northern Areas seems unlikely" until the gap between governmental control and resource utilization by mountain farmers will be resolved.

3.6 Development of Animal Husbandry as Perceived by Mountain Farmers

The changing socio-economic conditions, constraints, and opportunities to earn a living are reflected by the mountain population's reactions and perceptions. With regard to the management of natural resources and animal husbandry the grazing control during the vegetation period in summer as well as the control over the granted rights of the utilization of pastures and forests are important issues for individuals and the entire village communities. Conflicts within the villages or between neighbouring villages frequently occur, but these are mostly solved by local, informal institutions or by elected political representatives. Additionally, the communities also demand their direct participation in the sustainable management of the natural resources on local level in order to ensure their agro-pastoral strategies and subsistence.

The villagers' own perceptions of wealth indicators and disparities within the communities and of development demands also suggest that they are aware of the recent transitional processes. The integration of non-agrarian aspects into the range of indicators and felt needs on village-level proves that the people are realizing the new opportunities and challenges and started to actively improve their livelihood. As part of a "participatory appraisal" the people in the village of Hilbiche in Astor ranked all households by the means of a wealth differential, using their own indicators. They selected the acreage of cultivated area as the most important indicator, the households' herdsize, however, was only mentioned on rank four, after off-farm employment, and the level of education (cf. AHMAD & TETLAY 1994).

With regard to the translation of such perceptions into development activities, the people's own identification of priorities does not necessarily coincide with development programmes of governmental or non-governmental organizations. These are often confined to selected sectoral activities and technical interventions. Based on their existing knowledge of possible external support and programme incentives, the "participatory appraisal" among the population of Hilbiche resulted in the incorporation of the livestock sector into a list of "prioritised needs"[23]. In the next step, schemes and direct management interventions out of the different programme packages of AKRSP – which are confined to the increase of the agrarian productivity – were selected. (AHMAD & TETLAY 1994: 19). The expression and prioritisation of demands is obviously based on the respondents' knowledge of possible support and their contextual objectives. Thus, the local people are acting as subjects and try to gain profits from the available facilities and resources in order to improve and diversify their subsistence. Other elements of the farmers' indigenous expertise and differing demands are hardly adressed by these programmes.

### 3.7 Changes of Fodder Production

The manifold integration of agriculture and animal husbandry also includes fodder cultivation, mostly on inferior land. Due to the decrease of horses after the partition of Kashmir the acreage of lucerne declined also. Only recently, farmers have started again to sow lucerne at least on newly developed land. Additionally, the edges of fields and land not suitable for wheat or other crops are irrigated occasionally to improve the fodder supplies. According to MALIK (1996), up to nearly 40 percent of the cropped area is sown with fodder crops. Other figures for Astor and the "Northern Areas" in general estimate the percentage of lucerne and clover at 20 percent (WARDEH 1989: 4, 13, PILARDEAUX 1995: 95).

Similar to the lacking self-sufficiency with regard to grains, there is also a fodder production gap which is estimated for the entire "Northern Areas" at 26 percent (STREEFLAND, KHAN & v. LIESHOUT 1995: 85). WARDEH (1989: 7) further differentiates the gap of feedstuff: approximately 26 % of dry matter, 48 % of utilizable energy content, and 49 % of raw protein. The negative balance between the feedstuff quantities and contents also indicates the poor quality of the fodder. Since fodder for the winter season is the main constraint of animal

---

23 "Prioritised needs" are: irrigation water supply, afforestation, a local hospital; drinking water, livestock (i.e., improved breeds, vaccination, specialists' training etc.) (AHMAD & TETLAY 1994: 19).

husbandry, hay is also collected from adjoining slopes, although crop residues provide the major source of winter fodder. Nowadays, the households can rely on regular supplies of grain and flour from "down country", and the farmers changed the cropping patterns in favour of wheat and maize. These crops yield more straw than barley, which offers a shorter ripening period. Today the varieties are selected according to the people's preference for the taste of flour and because of the amount of straw yielded.

The fodder situation in Astor is less problematic since the natural pastures provide more fodder compared to the more arid conditions further north. According to an economic survey by MALIK (1996), the mean household in Astor provides 90 percent of the feedstuff inputs by its own production. Parts of the deficit are filled by pruning broad leafed trees or by the collection of hay from uncultivated land, which obviously is excluded from this survey. Several farmers, at least in the Rupal Gah, also feed parts of their grain yields, mostly barley, to their animals in winter. This is also practised in other villages as an emergency strategy in spring, especially to strengthen the oxen for ploughing. Grain, mainly intended for human consumption, can be easily substituted by purchased wheat and flour. For fodder, there is, however, only a very limited supply of surplus which is marketed between neighbouring families or villages.[24]

Since the volume of fodder significantly limits its transportation, most stocks of hay and straw, which are collected during summer and autumn in the summer field settlements, are fed to the animals during repeated visits in winter (NÜSSER & CLEMENS 1996a: 127). Recently, however, fewer families practice these winter visits (see above).

## 3.8 Mortality of Livestock

The repeatedly claimed low productivity of the local herds is also proved by historical and recent reports with regard to diseases. Between 1899/1900 and 1914/15, harvest damages and subsequent animal losses due to untimely or excessive rains or to the early setting of cold weather and snowfall occurred eight times, at least in parts of Astor (SINGH 1917: 64). In 1928 heavy losses of animals in Astor were reported because of "Rinderpest" together with "unusual rains" (IOR 1928).

Recent reports for Astor indicate mortality ratios up to one third of the stocks,[25] and for 1994, seven out of 15 villages reported animal diseases. Vacci-

---

24  MALIK (1996) reports a mean marketing ratio of the feedstuff demand of 7 % which has to be purchased; for fodder sales the reported marketing ratio is even zero in all villages. According to the authors' own experiences in the Rupal Gah, several farmers use to buy hay and straw. The prices increase from winter to spring. PILARDEAUX (1995: 98) states that hay has a growing importance as an intra-regional cash crop.

25  Cf. BHATTI & TETLAY (1995), MALIK (1996). NÜSSER (1998a: 115) also discusses the mortality and explains the high ratios by the "mal-feeding" practices, leading to losses of weight, low fertility, and growth, especially of pregnant animals and young stock, which are born mostly in spring during the fodder-gap-peak.

nation and other preventive measures are still insufficient in Astor. In 1992 42 percent even of the vaccinated or medically treated animals died, due to untimely or improper treatment (BHATTI & TETLAY 1995: 19). In 1994 a mean of only 38 Rs. was annually spent per household for animal vaccines and medicine (MALIK 1996), compared with the minimum daily wage of occasional labourers of Rs. 35 (for 1992, cf. DITTRICH 1995: 246). Constraints lay especially in the absence of veterinary facilities in the remote villages and in the farmers' unawareness of higher returns from smaller but healthy herds. Well planned and managed programmes of "vaccination reduced animal mortality 2.6 times, financial losses fivefold, and disease incidence sevenfold" in the Gilgit region (World Bank 1990: 52 based on AKRSP data). AKRSP has recently tried to cope with this problem in Astor as well. In co-operation with the governmental line departments training programmes for "village specialists" and improved supply structures for vaccines and medicine are developed.

HANSEN (1996: 57), however, attributes the problem of animal mortality also to the declining knowledge of traditional medicines and treatment. The development activities in Astor have discouraged the indigenous knowledge, although there are marketing potentials of herbs and medicinal plants, which were exploited more intensively in history (HANSEN 1996: 50–52). Only recently, AKRSP and IUCN have started activities for the survey and promotion of medicinal plants in several villages of Astor after 1995.

With the objective of increased productivity, WARDEH (1989: 11) argues in favour of culling strategies to reduce the overaged structure of the herds and to buy more and better feedstuffs for the remaining animals. This strategy has not been adopted by the farmers until recently, since their herdsize is usually to small for this strategy, and "... farmers are reluctant to reduce their herd and flock size." (World Bank 1990: 52).

### 3.9 Marketing of Livestock

Animal husbandry still is a key element of the "mixed mountain agriculture" practised around Nanga Parbat: 35 percent of the total gross farm income in Astor derives from animal husbandry, ranging from 20 to 47 percent (MALIK 1996). These figures include the cash equivalents of meat, milk, wool, and hides of the animals as well as their farm manure production. The majority of these products are, however, consumed within the households and the real cash income from animal husbandry is limited to approximately five percent of its gross output (MALIK 1996), ranging up to 15 percent in some villages. Until recently, the household consumption of meat and livestock products has been very low, and also in subsistence households, meat consumption is mostly limited to the winter season or to special events and festivals. The bulk of food consists of cereals and vegetables, which are supplemented by milk products and seldom by meat.

Marketing of animals and livestock products is mostly confined to the individual villages. Debtors often repay their credits in kind, e.g. animals, or butter and *ghi*, to local shopkeepers. Another minor marketing aspect is the sale of skins - together with scrap metal and used plastic items - to itinerant Pathan traders, who travel from village to village in late autumn. In autumn, itinerant traders from neighbouring valleys also travel from village to village to buy young stocks, which are sold to other farmers for breeding purposes or to the army for their meat. Other entrepreneurs buy animals in Azad Kashmir in September or October, cross over the Shontar Pass into the Mir Malik Gah and proceed further via the Deosai Plateau to Skardu where the animals are eventually sold to the army. Occasionally, local farmers also sell animals to these traders.[26]

For the village of Nagai, close to the "Line of Control", the local army garrison serves as an important marketing incentive. In this village eleven percent of the gross livestock outputs and 34 percent of the gross poultry outputs are marketed (MALIK 1996). All over Astor, the marketing potentials are, however, not fully developed, and formal marketing strategies and facilities are still lacking.

Similar to the national situation, also the "Northern Areas" of Pakistan are facing gaps in the supply of meat. In spite of the traditions and potentials of animal husbandry in this region, the demand gap is met by subsidized "imports" of live animals from "down country", smuggled animals, or the initiative of local traders, importing animals also from China.[27] The intra-regional marketing potentials of the livestock production can be evaluated from the transport of animal products via the "Karakoram Highway" to the districts of Gilgit and Baltistan: one third of all imports to Gilgit consists of animal products, with 15 percent for beef and mutton, nine percent for poultry, and nearly nine percent for milk products (KHAN & KHAN 1992: 15). Traders also bring buffaloes with trucks from "down country" to Jaglot, from where the animals are brought to Astor on foot to be slaughtered and sold in the local bazaar.

Thus, incorporated into an integrated approach, livestock rearing might offer additional potentials for cash income in the "Northern Areas" and also around Nanga Parbat. This requires, however, external assistance to develop reliable structures of input supplies and output marketing, so that the local farmers and stockholders gain economic incentives and can change their agro-pastoral strategies according to quality objectives. The current policy of subsidies discourages such changes.

---

26 Information kindly provided by R. Göhlen, who conducted studies in the Mir Malik Gah.
27 The direction of the commercial livestock trade towards the mountains, and not vice-versa, might be explained by the growing demand of the army and by the local population as well as by tourists in summer and, of course, by the lack of marketing structures for the local livestock production. Government subsidies for grain and livestock-transportation are discouraging the local production (World Bank 1990: 95). KREUTZMANN (1993a: 30) also reports that the number of places for slaughtering buffaloes in the bazaar of Gilgit increased from two (1985) to 15 (1990).

## 4. CONCLUSIONS

Although the stocking density of the alpine pastures around Nanga Parbat has increased during the past decades because of the growth of the animal herds, the agro-pastoral economy still has a sustainable ecological potential under the current socio-economic conditions. However, more and more families do not longer participate fully in the seasonal vertical migration with their herds. In a medium to long-term time-scale the male labour-force might decrease significantly due to the increasing demands for and opportunities of off-farm-employment. This will induce more intense changes within the management structures of the local agro-pastoral economies.

Up to the present the agro-pastoral management in the Nanga Parbat area is still practised in a subsistence manner. This must not be regarded as a purely backward strategy, since the entire household economies are no longer solely agrarian or subsistence economies. In a summarizing statement, STREEFLAND, KHAN & V. LIESHOUT (1995: 93, 114) conclude that agriculture in the "Northern Areas" is generally reduced to a part-time activity in terms of income and labour requirements. Most households, however, still rely on the utilization of agro-pastoral resources to meet their basic requirements. Rough estimates for Astor refer to the dichotomy of subsistence and market-oriented economies, and reveal that "... 90% of the farming population is in a late stage of subsistence ..." and "ten percent ... entered the early transitional stage" (STREEFLAND, KHAN & V. LIESHOUT 1995: 98-99). This simplified explanation does not fully reflect the actual situation, since elements of subsistence farming together with animal rearing are just a few components of the households' entire reproduction activities, which are focused on widely spreading the risks of income generation.

Simultaneous to the increasing integration of the mountain villages into the cash economy, additional marketing incentives for crop- and livestock production should be developed, especially in the proximity of the bazaars of Gilgit or Chilas and along the "Karakoram Highway". The local and regional ecology and economy provides niches, which might be "exploited" with external assistance or external supplies, at least in a medium-term perspective. This will offer potentials for certain farmers to specialize in particular aspects of mountain agriculture, like dairy farming, meat production, or fodder cultivation, and thus inducing intensified intra-regional exchange relations.

Until the early 1960s there are only few indications of intensive market-oriented transitional processes around Nanga Parbat (PILARDEAUX 1995: 104, after STALEY 1966). Astor was previously, however, not "isolated" from supra-regional exchange relations (cf. Chapt. 3.2). Some major indicators for the recent dimensions of change are the "modern" transportation facilities and the significant increase of "commodification", based on external supplies, especially of wheat, discouraging the local staple food production. Beside this, seasonal migration and the agrarian mechanisation induced socio-economic changes within the villages. With these changes and the increasing unbalanced exchange relations also the dependence from external entrepreneurs, either in economics or in politics, increased and thus the region's socio-economic vulnerability.

Indications of transition and change from the perspective of animal husbandry and agro-pastoral management strategies in the Nanga Parbat area are, however, not homogenous. Opportunities and impacts vary significantly. Apart from the settlements' varying access and distance to markets, there are also differences with regard to the settlement history and aspects of utilization rights. The socio-economic differentiation between and within villages increases as well. This transition particularly favours men, who actively utilize the opportunities of education and employment. Women are still confined to their homesteads and villages (GÖHLEN 1997), and often they have to face an even larger workload. Wealthier families with regular off-farm income delegate more and more agrarian and pastoral tasks to relatives, share holders, or even tenants and are thus transforming the social structures. Until recently, this has not led to definite changes of the "traditional" patterns of animal husbandry and agro-pastoral management strategies in this area. The socio-economic and political dynamics will certainly lead to further changes. Their direction and intensity, especially with regard to the ecological potentials of rangelands and high pasture resources, depend largely on the political conditions and on processes in the Pakistani lowlands.

## 5. REFERENCES

AHMAD, R. & K.A. TETLAY (1994): Planning for change: partnership at the grassroots level. Participatory planning and appraisal excercise with the VO and WO in Helbich, Astore. Gilgit: AKRSP (unpublished).

AKRSP/Aga Khan Rural Support Programme (1997a): Fourteenth annual review, 1996. Gilgit.

– (1997b): Ninth six-monthly progress report on AKRSP's expansion into Astore Valley. (January-June 1997). Gilgit.

– (1997c): Participatory planning workshop, FMU Astore, 27-28 March 1997. Gilgit: AKRSP (unpublished).

ALLAN, N.J.R. (1986): Accessibility and altitudinal zonation models of mountains. In: Mountain Research and Development 6 (3), 185–194. Discussion, comments, and reply: 6 (3), 195–206.

BAMZAI, P.N.K. (1987): Socio-economic history of Kashmir (1846–1925). New Delhi.

BHATTI, M.H. & K.A. TETLAY (1995): Socio-economic survey of Astore Valley. A benchmark survey. Gilgit: AKRSP (unpublished).

CASIMIR, M.J. & A. RAO (1985): Vertical control in the Western Himalaya: Some notes on the pastoral ecology of the nomadic Bakrwal of Jammu and Kashmir. In: Mountain Research and Development 5 (3), 221–232.

CLEMENS, J. (1999): Rural development in northern Pakistan. Impacts of the Aga Khan Rural Support Programme. In: DITTMANN, A. (Ed.): Mountain societies in transition. Contributions to the cultural geography of the Karakorum. Köln (= Culture Area Karakorum, Scientific Studies 6) (in press).

CLEMENS, J. & M. NÜSSER (1994): Mobile Tierhaltung und Naturraumausstattung im Rupal-Tal des Nanga Parbat (Nordwesthimalaja): Almwirtschaft und sozioökonomischer Wandel. In: Petermanns Geographische Mitteilungen 138 (6), 371–387.

– (1996): Gefahr für die Märchenwiese gebannt? In: Südasien 16 (2–3), 84–87.

– (1997): Resource management in Rupal Valley, Northern Pakistan: The utilization of forests and pastures in the area of Nanga Parbat. In: STELLRECHT, I. & M. WINIGER (Eds.):

Perspectives on history and change in the Karakorum, Hindukush, and Himalaya. Köln (= Culture Area Karakorum, Scientific Studies 3), 235–263.
DANI, A.H. (1989): History of Northern Areas of Pakistan. Islamabad.
DITTRICH, C. (1995): Ernährungssicherung und Entwicklung in Nordpakistan. Saarbrücken (= Freiburg Studies in Development Geography 11).
- (1997): Food security and vulnerability to food crisis in the Northern Areas of Pakistan. In: STELLRECHT, I. & M. WINIGER (Eds.): Perspectives on history and change in the Karakorum, Hindukush, and Himalaya. Köln (= Culture Area Karakorum Scientific Studies 3), 23–42.
- (1998): High-mountain food systems in transition – food security, social vulnerability and development in Northern Pakistan. In: STELLRECHT, I. & H.-G. BOHLE (Eds.): Transformation of social and economic relationships in Northern Pakistan. Köln (= Culture Area Karakoram Scientific Studies 5), 231–354.
DREW, F. (1875): The Jumoo and Kashmir territorries. A geographical account. London [Reprint: Karachi 1980].
EHLERS, E. (1995): Die Organisation von Raum und Zeit - Bevölkerungswachstum, Ressourcenmanagement und angepaßte Landnutzung im Bagrot/Karakorum. In: Petermanns Geographische Mitteilungen 15 (2), 105–120.
FREMEREY, M., KIEVELITZ, U. & Z. MAHAL (1995): Aga Khan Rural Support Programme Pakistan, Astore project. Project progress review, phase I (1992–95). Eschborn: GTZ (unpublished).
Gazetteer of Kashmir and Ladak; together with routes in the territories of the Maharaja of Jamu and Kashmir. Calcutta, 1890 [Reprint: Lahore, 1991].
General Staff India (Ed.) (1928): Military report and gazetteer of the Gilgit Agency. Simla.
GÖHLEN, R. (1997): Mobilität und Entscheidungsfreiraum von Frauen und Männern im Astor-Tal (Westhimalaya). In: VÖLGER, G. (Ed.): SIE und ER. Frauenmacht und Männerherrschaft im Kulturvergleich. Köln (= Ethnologica NF 22), 287–300.
Government of Pakistan, Population Census Organisation (Ed.)(1984): District census report of Diamir. Islamaband.
GREEN, M.J.B. (1993): Nature reserves of the Himalaya and the mountains of Central Asia. New Delhi.
GRÖTZBACH, E. (1993): Tourismus und Umwelt in den Gebirgen Nordpakistans. In: SCHWEINFURTH, U. (Ed.): Neue Forschungen im Himalaya. Stuttgart (= Erdkundliches Wissen 112), 99–112.
HANSEN, R. (1996): "Gene für den Norden – Pillen für den Süden": Heilpflanzen-Forschung in Astor (NW-Himalaya, Pakistan): Ausbeutung oder Förderung eines endogenen Entwicklungspotentials? In: Entwicklungsethnologie 5, 43–69.
- (1997): Remembering hazards as "coping strategy": Local perception of the disastrous snowfalls and rainfall of September 1992 in Astor Valley, Northwestern Himalaya. In: STELLRECHT, I. & M. WINIGER (Eds.): Perspectives on history and change in the Karakorum, Hindukush, and Himalaya. Köln (= Culture Area Karakorum Scientific Studies 3), 361–377.
HERBERS, H. (1998): Arbeit und Ernährung in Yasin. Aspekte des Produktions-Reproduktions-Zusammenhangs in einem Hochgebirgstal Nordpakistans. Stuttgart (= Erdkundliches Wissen 123).
IOR/India Office Library & Records. Administration Report of the Gilgit Agency for the year 1928. P&S/12/3288. London.
IUCN/The World Conservation Union (n.d.; 1998): Maintaining biodiversity in Pakistan with rural community development. Islamabad (= Annual Progress Report, January-December, 1997).
KHAN, F.U. (1991): A report on livestock in the Northern Areas. Gilgit: AKRSP (unpublished).
KHAN, M.H. & S.S. KHAN (1992): Rural change in the third world. Pakistan and the Aga Khan Rural Support Programme. New York etc. (= Contributions in Economics and Economic History 129).

Kitab Hukuk-e-Deh (1915/16). Note status of villages in Rupal Valley. Astor. (unpublished report, Revenue Office, Astor).
KREUTZMANN, H. (1989): Hunza. Ländliche Entwicklung im Karakorum. Berlin (= Abhandlungen Anthropogeographie 44).
- (1991): The Karakoram Highway: The impact of road construction on mountain societies. In: Modern Asian Studies 25 (4), 711–736.
- (1993a): Challenge and response in the Karakoram: Socio-economic transformation in Hunza, Northern Areas, Pakistan. In: Mountain Research and Development 13 (1), 19–39.
- (1993b): Development trends in the high mountain regions of the Indian subcontinent. In: Applied Geography and Development 42, 39–59.
- (1995): Globalization, spatial integration, and sustainable development in Northern Pakistan. In: Mountain Research and Development 15 (3), 213–227.
- (1996): Tourism in the Hindukush-Karakoram: A case study on the valley of Hunza (Northern Area of Pakistan). In: BASHIR, E. & ISRAR-UD-DIN (Eds.): Proceedings of the Second International Hindukush Cultural Conference. Karachi (= Hindukush and Karakoram Studies 1), 427–437.
LAWRENCE, W. (1875): The Valley of Kashmir. [Reprint: Srinagar, 1967; Mirpur (Azad Kashmir), not dated].
- (Ed.) (1908): The Imperial Gazetter of India. Kashmir and Jammu. Calcutta (= Provincial Series 13). [Reprint: Lahore 1983].
MALIK, D. (1996): Farm household income & expenditure survey 1994: Socio-economic profiles of the 15 villages surveyed in Astor. Gilgit: AKRSP (unpublished).
MILLER, D. (1997): Range management and pastoralism. In: ICIMOD Newsletter 27, spring 1997, 4–6.
MOCK, J. (1995): Karakoram notes. Objects of desire in the Northern Areas. In: Himal, March/April 1995, 19.
NÜSSER, M. (1998a): Nanga Parbat (NW-Hiamalya): Naturräumliche Ressourcenausstattung und humanökologische Gefügemuster der Landnutzung. Bonn (= Bonner Geographische Abhandlungen 97).
- (1998b): Animal husbandry and fodder requirements around Nanga Parbat, Northern Areas, Pakistan: Recent and historical perspectives of human-ecological relationships. In: STELLRECHT, I. (Ed.): Karakorum-Hindukush-Himalaya: Dynamics of Change. Köln (= Culture Area Karakorum, Scientific Studies 4, pt. I), 319–337
NÜSSER, M. & J. CLEMENS (1996a): Impacts on mixed mountain agriculture in the Rupal Valley, Nanga Parbat, Northern Pakistan. In: Mountain Research and Development 16 (2), 117–133.
- (1996b): Landnutzungsmuster am Nanga Parbat: Genese und rezente Entwicklungsdynamik. In: KICK, W. (Ed.): Forschung am Nanga Parbat. Geschichte und Ergebnisse. Berlin (= Beiträge und Materialien zur Regionalen Geographie 8), 157–176.
PARVEZ, S. & Ehsan-ul-Haq JAN (n.d.): Growth, poverty and inequality in the programme area. A comparison of 1991 and 1994 results based on FHIES data. Summary of some preliminary results. Gilgit: AKRSP (unpublished).
PILARDEAUX, B. (1995): Innovation und Entwicklung in Nordpakistan. Über die Rolle von exogenen Agrarinnovationen im Entwicklungsprozeß einer peripheren Hochgebirgsregion. Saarbrücken (= Freiburg Studies in Development Geography 7).
- (1997): Agrarian transformation in Northern Pakistan and the political economy of highland-lowland interaction. In: STELLRECHT, I. & M. WINIGER (Eds.): Perspectives on history and change in the Karakorum, Hindukush, and Himalaya. Köln (= Culture Area Karakorum Scientific Studies 3), 43–57.
- (1998): Surviving as a mountain peasant - innovation, development and the dynamics of global change in a high-mountain region. In: STELLRECHT, I. & H.-G. BOHLE (Eds.): Transformation of social and economic relationships in Northern Pakistan. Köln (= Culture Area Karakorum Scientific Studies 5), 355–426.

Revenue Office, Astor (Ed.)(1990): Census of housing list as of Nov. 1990, Astor subdivision. (unpublished).
Revenue Office, Chilas (Ed.)(1981): Village-wise housing and population census 1981, Chilas subdivision. (unpublished).
– (1990): Village-wise housing and population census 1990, Chilas subdivision. (unpublished).
RHOADES, R.E. & S.E. THOMPSON (1975): Adaptive strategies in alpine environments: beyond ecological particularism. In: American Ethnologist 2, 535–551.
SCHOLZ, F. (1982): Einführung. In: SCHOLZ, F. & J. JANZEN (Eds.): Nomadismus - Ein Entwicklungsproblem?. Berlin (= Abhandlungen Anthropogeographie 33).
– (1991): Von der Notwendigkeit, gerade heute über Nomaden und Nomadismus nachzudenken. In: SCHOLZ, F. (Ed.): Nomaden, mobile Tierhaltung. Berlin, 7–37.
SINGH, T. (1917): Assessment Report of the Gilgit Tahsil. Lahore.
SNOY, P. (1993): Alpwirtschaft im Hindukusch und Karakorum. In: SCHWEINFURTH, U. (Ed.): Neue Forschungen im Himalaya. Stuttgart (= Erdkundliches Wissen 112), 49–73.
STELLRECHT, I. (1997): Dynamics of highland-lowland interaction in Northern Pakistan since the 19th century. In: STELLRECHT I. & M. WINIGER (Eds.): Perspectives on history and change in the Karakorum, Hindukush, and Himalaya. Köln (= Culture Area Karakorum, Scientific Studies 3), 3–22.
STÖBER, G. (1993): Bäuerliche Hauswirtschaft in Yasin (Northern Areas of Pakistan). Bonn (unpublished report to the 'Deutsche Forschungsgemeinschaft').
STREEFLAND, P.H., KHAN, S.H. & O. VAN LIESHOUT (1995): A contextual study of the Northern Areas and Chitral. Gilgit.
TROLL, C. (1939): Das Pflanzenkleid des Nanga Parbat. Begleitworte zur Vegetationskarte der Nanga Parbat-Gruppe (Nordwest-Himalaja) 1:50.000. Leipzig. (= Wissenschaftliche Veröffentlichungen des Deutschen Museums für Länderkunde zu Leipzig, N. F. 7), 149–193.
UHLIG, H. (1995): Persistence and change in high mountain agricultural systems. In: Mountain Research and Development 15 (3), 199–212.
WARDEH, M.F. (1989): Livestock production and feed resources in Gilgit District of Northern areas of Pakistan. A preliminary study. Gilgit (= AKRSP Livestock Paper 4).
World Bank (1990): The Aga Khan Rural Support Program. Second interim evaluation. Washington (= A World Bank Operations Evaluation Study).
– (1995): Pakistan. The Aga Khan Rural Support Program. A third evaluation. Washington (= Report No. 15157-PAK).
YOUNGHUSBAND, F. (1911): Kashmir. London [Reprint: Mirpur (Azad Kashmir) not dated].

Tab. 6: Major events regarding animal husbandry and pastoral strategies around Nanga Parbat.

a. Astor

| Time-Scale approx. | Event and Actors | Implications |
|---|---|---|
| pre 1851 | "Chilasi raids" throughout Astor | robberies of animals, properties, and people as slaves by Chilasi coming across the passes, abandonment of several villages and pastures |
| pre 1894 | *kar begar / res* system | forced labour for the villagers along the transportation routes, additional burden, lack of male workforce, often withdrawal of the families to higher altitudes |
| 1894 | "Gilgit-Bandipur-Road": mule track, from Kashmir via Astor and Bunji to Gilgit, built by British and Kashmiri authorities | income opportunities for local owners of pack animals, and from the sale of fodder |
| colonial period | land revenue, some support and innovations | taxes to be paid in cash, occasionally also in kind, partial introduction of improved seeds, credits for pack animals |
| 1925 | "Gilgit-Bandipur-Road": enlarged for 2 mules side by side | see above |
| 1932–39 | first scientific expeditions to the Nanga Parbat | engagement of porters and eventually also of mountain guides |
| 1935 | "Gilgit Lease", Kashmiri territories west of the Indus leased by the British | Astor remained with the *State of Jammu and Kashmir*, the British increased traffic via the Babusar road (see Tab. 6.b), Astor's partial loss of the transit function |
| 1947 | partition of British India and Kashmir, "Case Fin Line"/ "Line of Control" (LoC) | total loss of the transit function, peripheralisation of Astor |
| after 1948 | among others: location of Pakistani Army | establishment of new supply lines, from the Indus to LoC, income opportunities within the army, increase of the no. of cattle, after the abolition of the Hindu-ban of meat consumption; nomadic Bakrwal groups come to high pastures of Astor |
| 1953 onwards | further scientific expeditions to the Nanga Parbat | engagement of porters, and eventually also of mountain guides |
| 1959–65 | "Indus Valley Road" | jeepable track along the Indus up to Gilgit, better supplies with commodities due to reduced transportation costs |
| late 1950s | first jeep roads in Astor, especially for army supplies | improved access and supplies with commodities |
| ca. 1960 | start of labour migration | beginning transition of subsistence economies |
| 1972–74 | abolition of land taxes, introduction of 'civil supply' system | increased commodities' supplies, changed cropping systems, stop of farmers' caravans to Azad Kashmir, decline of horses |
| 1975 | establishment of the "Astor Wildlife Sanctuary" | no impacts on the agro-pastoral management |

| 1978 | "Karakoram Highway" | metalled and truckable road, along the Indus up to Gilgit and further to China; improved access and supplies |
| 1985–86 | jeep track to Churit & Tarishing | improved access and supplies, also for tourists |
| 1986 | start of agrarian mechanisation by private entrepreneurs | reduction of labour-peaks, however, no significant decrease of draught animals |
| 1993 | start of the "Aga Khan Rural Support Programme" | support in agriculture, animal husbandry, natural resource management, income generation, also for women |
| 1995 | declaration of "Fairy Meadows National Park" and "Deosai Wilderness Park" | restrictions of agro-pastoral and forest utilization for the local population feared |
| 1997 | start of IUCN's biodiversity project | conservation activities and "village management plans" of natural resources, for Bunji, Doian, Dashkin and Turbaling |

Source: compiled from different publications, cf. references.

b. Raikot

| Time-Scale *approx.* | Event and Actors | Implications |
|---|---|---|
| pre 1880 | settlement of Gujurs | differentiation/stratification of the population, former nomads as tenants and/or shepherds |
| | *karbegar / res*-system *bar rajaki* | forced labour for the villagers along the colonial transportation routes, additional burden, lack of male workforce, often withdrawal of the families to higher altitudes, population of Gor was later freed from this system |
| 1896 | "Gunar Farm", governmental fodder supply, later (1920) "Durbar Cultivation Farms", at Bunji, Pari, Chilas | improvement of colonial transportation system, intensified use of pack animals, expansion into seed and crop testing and production, incl. fodder crops |
| 1898/ 1935 | mule tracks, via Babusar Pass | bypassing the Astor Valley, improvements under the British after the "Gilgit Lease" |
| 1920 | establishment of the "Transportation Department" | introduction of governmental pack animals |
| 1932–39 | first scientific expeditions to Nanga Parbat | engagement of porters, and eventually also of mountain guides |
| 1947 | partition of British India and Kashmir | |
| 1948/49 | Kaghan Route, via Babusar Pass | first jeepable track towards Gilgit; bypassing the closed Astor Valley, improved access and supplies |
| 1953 | first climbing of the summit | further expeditions |
| 1959–65 | "Indus Valley Road" | jeepable track along the Indus up to Gilgit, better supplies with commodities due to reduced transportation costs |
| 1978 | "Karakoram Highway" | metalled and truckable road, along the Indus up to Gilgit and further to China; improved access and supplies |
| 1988/ | jeep track to Tato, | private project by an absent entrepreneur, combined |

| | | |
|---|---|---|
| 1991 | and further to "Fairy Meadows" | with forest extraction and plan of hotel construction, extension to Fairy Meadows destroyed by debris flows, intensification/commerialisation of tourism |
| 1992/93 | "Raikot Serai" by local entrepreneur | private camping ground at "Fairy Meadows" after resisting a commercial hotel complex, further tourism activities by other local families |
| ca. 1994 | "Himalayan Wildlife Project" | local Non Governmental Organization in Gor, Raikot |
| 1995 | declaration of the "Fairy Meadows National Park" | restrictions of agro-pastoral and forest utilization for the local population feared |
| 1998 | start of IFAD's rural development project | broad based support in agriculture, animal husbandry, natural resource management, income generation |

Source: compiled from different publications, cf. references.

# WHY ARE MOUNTAIN FARMERS VEGETARIANS? NUTRITIONAL AND NON-NUTRITIONAL DIMENSIONS OF ANIMAL HUSBANDRY IN HIGH ASIA

HILTRUD HERBERS

## 1. INTRODUCTION

Animal husbandry is a substantial element of the high mountain economy in the Hindukush, Karakorum and Himalaya. At the highest altitudes of the permanent settlement area, where it becomes increasingly difficult to grow cereals due to low temperatures and short summer seasons, animal husbandry is even more important than crop cultivation. Normally, such a specialisation of the production is clearly reflected in consumption patterns. In other words, keeping livestock often entails a diet rich in animal products and vice versa. This relationship is thought to hold true especially for communities which produce on a subsistence basis, i.e., mainly for the supply of their own households. In these cases, producers and consumers are, by and large, identical. Despite the obvious significance of keeping livestock, however, the local diet of many regions in High Asia proves to be predominantly vegetarian. Therefore, the correlation between production and consumption is not necessarily very strong, even if only few products are marketed. I will show in this paper why animal husbandry in High Asia is not primarily aimed at meeting nutritional needs and what consequences this has for the nutritional status of the people. During a two-year field research in the Yasin Valley, located in Northern Pakistan at an altitude of 2,160 2,710 m (see Fig. 7 in introducing chapter by EHLERS/KREUTZMANN in this volume), I examined the multiple functions which livestock fulfils in high mountain farming. Before presenting the results of my study, I will give a brief description of the study area.

Like other regions in High Asia, Yasin is doubly disadvantaged in that it is not only located in a developing country but also in a high mountain area. In social, economic and infrastructural terms it therefore takes up an extremely peripheral position. The approximately 30,000 inhabitants of the valley subsist mainly on agricultural products, which they cultivate in an extremely arid environment at an altitude allowing only a single crop. The agricultural yields are usually not sufficient to supply enough food for the family, which consists, on average, of ten to eleven members. Supplementary sources of income from off-farm activities help to fill the gap to various degrees. Since the people earn little more than they require for their most basic needs, their standard of living is relatively low.

Regarding basic services, such as school education, medical care and the provision of domestic commodities, the people of Yasin are largely restricted to local facilities. Gilgit, the commercial and administrative centre of the region, is located 120 km to the east. The poor condition of the roads and the insufficient public transport system prevent the people from travelling frequently to Gilgit and from making use of the various facilities in town. Therefore, governmental and non-governmental organisations try to increase the number of schools, health posts etc. in the valley. Since the majority of the inhabitants belong to the Islamic community of the Ismailiya, it is especially the Aga Khan Foundation which provides support. In spite of the help of this and other organisations, however, Yasin's infrastructure remains insufficient both in quantitative and qualitative terms.

## 2. AVAILABILITY OF ANIMAL PRODUCTS IN YASIN

The fact that all farmers of Yasin keep at least a few animals points to the importance of animal husbandry for subsistence. Yaks, cattle, goats, sheep, donkeys, horses and chicken are the main breeds reared in the valley. Other species such as buffaloes, which are widespread in the lowlands of Pakistan, do not breed in the cold climate of this altitude, while pigs, which could live in this environment, are not kept due to Muslim belief. To assess the number of animals per household, I carried out a livestock survey in the villages of Sandi and Barkulti, both located in the central part of the valley. This survey showed that chicken are most common, followed by goats, sheep and cattle (see Fig. 1). In addition, almost every household keeps at least one donkey, whereas only a few households rear some yaks or a horse, and a negligible number have ducks or rabbits. Except for chicken, the absolute number of animals per household did not change much from 1982/83 to 1992/93. The total number of all breeds, excluding chicken, increased only slightly from 13.8 to 14.0 in Sandi and from 16.7 to 18.4 in Barkulti during this decade.[1] Relative to the size of the households, the number of animals appears rather low.

The restricted availability of fodder in the valley is the main factor which prevents a further increase in the number of livestock. The small size of arable lands of the households (about 1–2 ha) are almost exclusively cultivated with cereals and other crops for human consumption. Lucerne (alfalfa), clover and other fodder plants are only grown on fields of minor quality, e.g. new land or areas located on steep slopes. Therefore, the animals are fed with grass from the fringes of the fields, tree leaves, weeds, straw, and sometimes food leftovers. In order to improve the fodder supply, the livestock is taken to the alpine meadows on the high pastures from March/April to October/November. Only one cow or goat remains in the homestead to provide the family with milk. However, not all households have access to the high pastures, because they either lack property

---

1 The total number of breeds in 1992/93 ranged from 0 to 62 in Sandi and from 2 to 51 in Barkulti. The only family in Sandi without animals other than poultry was a newly established household which had already ten chicken.

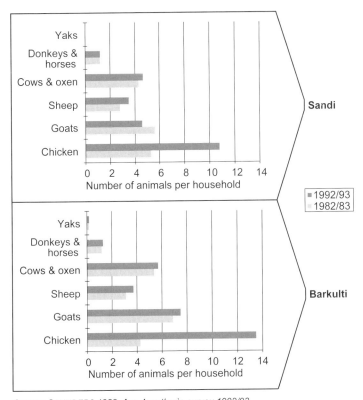

Fig. 1: Number of livestock in two villages in Yasin

and utilisation rights or labourers to tend the animals and process the milk in the summer villages.[2] In this case, the households try to put their livestock in the care of other farmers who do have access to the high pastures. For their services the latter receive a share of the milk products which they process from the milk of the owner's cows and goats.

Even if the animals spend the summer on the comparatively rich meadows of the high pastures, the forage remains deficient. Consequently, the output of milk and meat is rather low. A cow, for instance, gives barely two litres milk per day in summer and less than one litre in winter, and a hen lays only one or two eggs per week.

Until 1995, only the lower part of the Yasin Valley was supplied with electricity. Since then, a second hydroelectric power station has been put into operation

2  For details on property rights and labour organisation on the high pastures cf. STÖBER & HERBERS in this volume.

and today provides most of the upper valley with energy.[3] However, neither the improved energy supply nor the households' additional incomes allow families to have a refrigerator or a freezer and thus to store animal products for a longer time. Hence, animals are slaughtered only in winter, i.e., in November or December.[4] This meat is kept in a separate store room, which is not heated, and is consumed until February.[5] By contrast, milk is available especially during the main lactation period of the cows and goats in summer, i.e. in the hottest months of the year. Hence, milk, cottage cheese, cream or other fresh products have to be consumed directly. Only few dehydrated milk products, such as butter or *qurút* (solid protein concentrate) do not perish under the climatic conditions in Yasin.

## 3. THE VEGETARIAN DIET OF THE MOUNTAIN FARMERS IN YASIN

The *cuisine* of *Yasin* is characterized by a wide range of sweet and savoury dishes made of milk, meat, eggs, cereals, pulses, vegetables and fruits. Among the dishes made from animal products, *śerbát* (wheat flour roasted in rancid butter), *mul* (salted milk pudding with butter), *γulmandí* (a kind of pancake layered with curd or salted milk pudding) and *amestóno* (salted milk pudding with roasted linseeds and butter) are particularly popular. The same holds true for cooked or fried meat with rice. These dishes are essential elements of many festivities (Fig. 2), including wedding ceremonies, funerals and important religious and agricultural festivals, such as *qorbaní*, which is celebrated in remembrance of Ibrahim sacrificing his son to god,[6] or *thémišing* at the end of the agricultural season. However, the consumption of such dishes contributes only marginally to a balanced diet, since there are only one or two such feasts per month. In addition, not all festivities in the course of the year are celebrated with dishes containing animal products or other special ingredients (Fig. 2).

Apart from the various festivals, there are only two periods during which animal products are part of the farmers' diet. Slaughtering an animal at the end of the year is a special event for the whole family, which no one will miss (see Photo 1). Generally, a household kills one animal for the winter supply. This is usually a goat or a sheep, as only few households can afford to slaughter a cow. However, in the winter of 1991/92, 30% of the surveyed households in Sandi and 35% of those in Barkulti were not even able to spare a single goat or sheep for consumption. The comparatively low numbers of livestock and the fact that the animals

---

3   The side valleys Thui and Qorkulti as well as the settlements on the high pastures are still without electricity.
4   The custom of slaughtering animals in winter has the additional advantage that some of the limited fodder reserves are saved for other animals.
5   To kill an animal and store it for the winter is called *nasálu* all over Northern Pakistan. Literally it means "smelling meat", which points to the fact that meat is difficult to preserve in this area.
6   In other Islamic regions the festival is called ᶜ*id ul-adha*.

Why are Mountain Farmers Vegetarians? 193

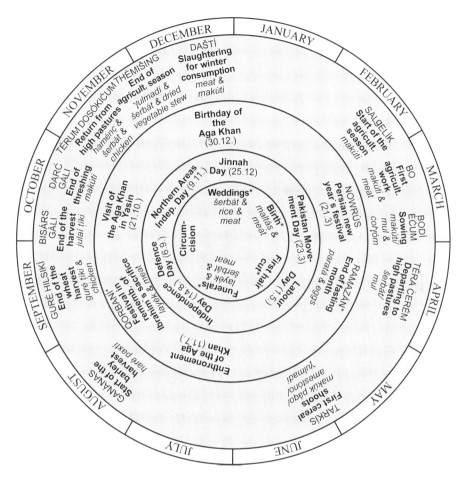

DARĆ GÁLI = Burushaski name of the festival
End of threshing = Reason of the festival
makúti = Name of the festival dish

| 1. Circle | Month |
|---|---|
| 2. Circle | Agricultural festivals |
| 3. Circle | Religious festivals |
| 4. Circle | State holidays |
| 5. Circle | Rites de passages |

& and   / or
* Varying dates

Source: Author's survey 1992/93
Design: H. Herbers   Graphics: J. Sehner

**Festival dishes:**

| | | |
|---|---|---|
| amestóno | = | Salted milk pudding with roasted linseeds and butter |
| ćoɫγóm | = | Noddle soup with meat |
| guré tíki | = | Thick leavened wheat bread |
| haménc | = | Kind of cottage cheese |
| haré paxtí | = | Barley porridge |
| julaí tíki | = | Thick leavened wheat bread with walnuts |
| layék | = | Wheat porridge with meat |
| makúk páqo | = | Sweet flat bread with melted butter |
| makúti | = | Sweet dish made of walnuts or apricot seeds |
| mul | = | Salted milk pudding with butter |
| paratá | = | Fried flat bread |
| šerbát | = | Wheat flour roasted in butter |
| γulmadí | = | Pancakes layered with curd or salted milk pudding and butter |

Fig. 2: Festivals and their concomitant food regulations

are needed for non-nutritional purposes[7] explain why these households do not slaughter their animals. The only meat which is eaten quite regularly throughout the year is poultry. It also requires special occasions to slaughter a chicken, but these are more frequent. Thus, a hen is often prepared if a close relative visits the family after a long time or if a household member recovers from a serious disease.

Photo 1: A cow is slaughtered for winter consumption (Sandi, November 1993)

In Yasin, cow and goat milk is mostly used to prepare tea. Black tea boiled with salt and milk is taken with bread two to four times a day, depending on the season of the year. While in summer the majority of the households in Yasin have usually enough milk to prepare tea, it becomes a scarce commodity in the winter season. Tea without milk, however, is not only considered as unpalatable; serving *troq ćay* to guests is also embarrassing for the host. To prevent such situations, households borrow milk from their neighbours or use sheep milk, which is usually not used for consumption.

As mentioned above, milk is more easily available during summer. This holds particularly true for the high pastures, where most of the livestock stays from March/April to October/November. In both localities, i.e. in the summer settlements and in the permanent villages, the surplus of milk is processed into various products. In this regard, the highest priority is not given to milk products

---

7   Cf. the discussion in section 4.

Why are Mountain Farmers Vegetarians?   195

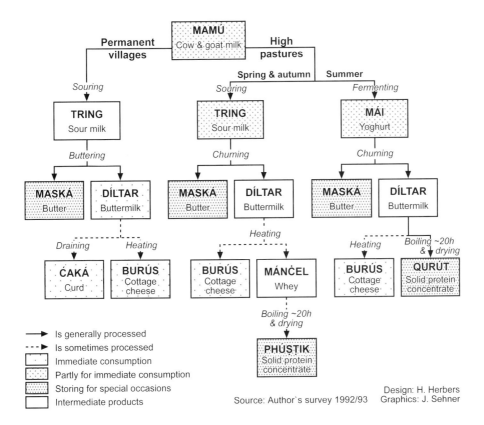

Fig. 3: Range of major milk products processed in Yasin

for immediate consumption, but to butter and *qurút*. While the former is essential for many festival meals (Fig. 2), the latter is necessary to prepare noodle soups, which are frequently eaten in winter, and to cure diseases, such as cold and influenza infections. Thus, out of the wide range of possible milk products,[8] only few are commonly prepared for every-day consumption (Fig. 3). On the high pastures, only sour milk and yoghurt are used for immediate consumption, while it is mainly buttermilk and sometimes a kind of cottage cheese (*burús*) or curd (*ćaká*) in the homesteads. Since it is only one female household member who spends the summer on the high pastures and who is responsible for all tasks there (e.g. housekeeping, food preparation, milk processing), it is the shortage of female labourers which prevents the processing of other milk products in this location. In the permanent villages, it is rather the scarcity of milk which compels the household members to consume predominantly buttermilk, a product, which is more filling and feeds more people than the processed milk products *burús* and

---

8   For the complete range of possible milk products, cf. HERBERS 1998, 78.

*ćaká*. Moreover, butter and *qurút* yield high profits on the markets in Gilgit. Many households depend on these earnings, since alternative sources of income are insufficient or missing. Thus, the availability of milk products is further limited, because only a certain portion of it is reserved for household consumption.

Compared to animal products, vegetables, fruits and pulses are easier to store. In Yasin, leafy vegetables, tomatoes, onions, beans, apricots and other garden products are dried, whereas potatoes, carrots, turnips, cabbages and pumpkins are kept over several months in holes in the ground or other cool places to be consumed in winter. However, since the plots of arable fields of the mountain farmers are small, plant products are scarce too. Hence, up to four meals per day consist exclusively of tea and different varieties of bread. Only the dinner contains a warm dish, which is usually a vegetable stew. Participant observations (Tab. 1) in the author's host family confirmed that in summer meat is rare, and milk products are served only occasionally. After slaughtering in winter, meat is more frequently available.

However, it is not the warm dish which constitutes the main component of the dinner. Chicken, for instance, is quite often served in the evenings (Tab.1), but portions are relatively small since most households do not slaughter more than one hen for a meal and the animal has to be divided among a large number of family members. Meat, milk products or vegetable stews are thus rather side dishes, which add flavour to the meals. Bread, predominantly made of wheat flour, is the major constituent of every meal and provides most of the daily energy intake for the household members.

Tab. 1: Dinner composition in a farmer's household in Yasin

| Day | May | July | December |
|---|---|---|---|
| 1 | Lentil soup, sweet dish | Stew of caper petals | – |
| 2 | Stew of fresh leaf veget. | Tea with bread | – |
| 3 | Stew of fresh leaf veget., buttermilk | Stew of fresh leaf veget. | – |
| 4 | Stew of fresh leaf veget. | Buttermilk, lettuce | – |
| 5 | Lentil soup | Stew of fresh leaf veget. | Rice, chicken |
| 6 | Stew of fresh leaf veget., wild duck | Stew of fresh leaf veget. | Stew of dried leaf veget. |
| 7 | – | – | Potato stew |
| 8 | Chicken soup | Stew of fresh leaf veget. | Pea soup |
| 9 | Stew of fresh leaf veget. | Stew of wild veget. | Potato stew, chicken, buttermilk |
| 10 | – | Stew of fresh leaf veget. | Goat meat |
| 11 | – | Stew of fresh leaf veget. | Stew of carrots and turnips |
| 12 | – | – | Noodle soup with milk |
| 13 | – | Chicken soup, lettuce | Potato stew, maize bread |
| 14 | – | Beef soup | Stew of dried leaf veget., goat meat |
| 15 | – | Stew of fresh leaf veget. | Carrot stew |

| | | | |
|---|---|---|---|
| 16 | – | – | Rice, goat meat |
| 17 | Stew of fresh leaf veget. | Buttermilk, lettuce | Pea soup |
| 18 | Pea soup | – | Stew of fresh beans, goat meat |
| 19 | Stew of fresh leaf veget. | Stew of fresh leaf veget. | – |
| 20 | Stew of fresh leaf veget. | Rice, lentil soup | Lentil soup |
| 21 | Rice, chicken, stew of fresh leaf veget. | – | Potato stew, goat meat |
| 22 | Stew of fresh leaf veget., walnut bread | – | Rice, goat meat |
| 23 | Stew of fresh leaf veget., buttermilk | – | Stew of dried leaf veget. |
| 24 | – | Buttermilk, lettuce | Potato stew, goat meat |
| 25 | Stew of fresh leaf veget. | Stew of fresh leaf veget. | Lentil stew |
| 26 | Stew of fresh leaf veget. | Stew of fresh leaf veget. | – |
| 27 | Stew of fresh leaf veget. | Buttermilk | Stew of carrots and turnips |
| 28 | Stew of fresh leaf veget. | Stew of fresh leaf veget. | Stew of dried leaf veget. |
| 29 | Stew of fresh leaf veget. | Stew of fresh leaf veget. | Lentil soup |
| 30 | Stew of fresh leaf veget. | Lentil soup | Cabbage stew |
| 31 | Stew of fresh leaf veget. | Stew of fresh leaf veget. | – |

| | | | |
|---|---|---|---|
| Freq. of meat consumption | | | |
| – totally | 3 | 2 | 9 |
| – weekly* | 1,0 | 0,6 | 2,6 |
| Freq. of milk & diary consumption | | | |
| – totally | 2 | 4 | 2 |
| – weekly* | 0,6 | 1,2 | 0,6 |

– No data available
* Weighted arithmetic mean
Source: Author's survey 1992/93

The intra-household food allocation, which reflects the social hierarchy of the community, gives preference to certain individuals. The best and largest portions of the meal are given to guests, especially when they are men or rare visitors, or when they hold prestigious positions (e.g. representatives of the Ismailiya community, physicians or teachers). Male household members come second when food is distributed. The older the man, the higher is his social standing. The same holds for women and children, who rank at the very end of the intra-household food allocation system. Thus, the latter profit only marginally from a dish with meat, butter or other highly esteemed foods. By way of example, the majority of the women interviewed mentioned that they hardly eat eggs, since these are exclusively reserved for the breakfast of male household members or for unexpected male guests. However, there are particular animal products of high nutritional value which are only consumed by women and children. The so-called *gušéngia xorák*, women's food, contains colostrum from cows and blood from goats and sheep.[9] According to the indigenous interpretation of Islamic instruc-

---

9   The colostrum, i.e. the milk of the first five days after calving, is processed to a kind of

tions, colostrum is classified as *makrú*, that means neither *harám* (forbidden) nor *halál* (allowed), and blood as *harám*.[10] Therefore, men refuse to eat these animal products in order not to commit a sin against Islam; it is tolerated, however, that women eat these inferior products.

Although the farmers in Yasin cultivate crops and keep livestock mainly for domestic purposes, they do not achieve self-sufficiency with respect to plant and animal products. Most families have to buy additional food on the market in order to cover their needs. These purchases, however, cannot supply all of the items needed. Local shopkeepers sell only the staple food wheat, tea, salt, sugar and other products which can not be grown in the valley, as well as a small range of non-food commodities. With very few exceptions, neither meat and milk products nor fresh vegetables and fruits are available in local shops.[11] To purchase these food items, people have to travel as far as Gilgit. Male household members who travel sometimes into town for other business take meat and fruits home with them if they can afford it. Incomes are low, however, and there is often not sufficient money to buy these items.

According to the memories of elderly people, hunting was an important source of meat in former times. Wild goats, especially *Capra ibex*, which were previously hunted in great numbers, are now almost extinct in Yasin. Thus, only small game, such as ducks and hares, are available. Yet, this irregular and small supply of meat is of little significance in nutritional terms.

To sum up, one can say that bread and tea are the central constituents of most meals in Yasin. Since animal products are consumed in small quantities and infrequently, the diet is more or less vegetarian. Due to the uneven intra-household food allocation, the diet of the privileged household members can be described as ovo-lacto-vegetarian, while that of the less privileged members is predominantly vegan.

Since comprehensive studies of nutritional issues in High Asia are rare, it is difficult to compare the situation in Yasin with that in other parts of this mountain region. There are, however, many studies on agriculture in the Hindukush, Karakorum and Himalaya, which touch on nutritional aspects. A survey of selected studies shows that the consumption patterns in other valleys in Northern Pakistan are largely comparable to those in Yasin. The diet in Chitral and Hunza, for instance, is as poor in meat, milk and dairy products as in Yasin (ALLAN 1990, 409; HASERODT 1989, 122–123; KREUTZMANN 1989, 146–147; MÜLLER-STELL-

---

cottage cheese (*haménç*), while blood is used to prepare a dish (*rank dóhalta*) with animal fat, wheat flour and salt. Food reserved for women includes also some dishes prepared after delivery or to cure menstrual pains. These dishes, however, are not specially made of animal products.

10 The Holy Koran and the instructions of the Prophet Muhammad distinguish only between two categories of food, i.e. allowed and forbidden. There are, however, many products, which Sunnis, Shias and Ismailis classify differently (LEWIS et al. 1965, 1068–1071).

11 The only dairy product available in the bazaar is milk powder, which some people purchase to prepare tea or to feed infants. Occasionally, one also finds a few eggs or a chicken, offered for sale by some private households.

RECHT 1979–80, 77). On the other hand, animal products, except meat, are more plentiful in regions which are hardly suitable for cultivation and which therefore leave the local population no other choice than to rely on animal husbandry. The same dietary habits can be found in communities whose members are traditionally mainly pastoralists, such as the Bakkarwal of Jammu and Kashmir or the Nurzay Durrani Pashtuns of Western Afghanistan (ATTENBOROUGH, ATTENBOROUGH & LEEDS 1994, 383–389; BISHOP 1992, 237; CASIMIR 1995, 91, 119; 1998: 48–54).[12] However, even under these circumstances the provision of animal products for consumption is only one of the various functions of keeping livestock. At least some of the functions which will be discussed below with respect to Yasin are likely to be the same in many other parts of High Asia.

## 4. NON-NUTRITIONAL DIMENSIONS OF ANIMAL HUSBANDRY IN YASIN

The analysis provided above reveals that in Yasin livestock is kept for a number of non-nutritional reasons related to the specific way of life in the high mountains of the Hindukush, Karakorum and Himalaya. Agricultural as well as economic, social and cultural factors determine the specific functions of animal husbandry in these areas.

### *Agricultural dimensions*

A highly diversified system of economic activities enables the inhabitants of Yasin to live in this extremely arid region.[13] Among the many activities which the people pursue to secure their livelihood, off-farm employment and agriculture are predominant, because they provide the largest share of the household's income. As the first aspect is not relevant to the argument presented here, I shall restrict myself to the agricultural activities.

The practice of combined mountain agriculture consists of two major components which are closely linked, i.e. the cultivation of land and animal husbandry. While the crops from the fields provide fodder (e.g. straw, weed) to feed the animals in winter, the latter supply manure for the fields to keep the soil fertile. Moreover, animals are necessary to carry out several agricultural tasks, such as

---

12 Moreover, locations with a high influx of migrants, such as governmental officials, soldiers or merchants, and of foreign tourists are usually better supplied with animal products. This can be observed especially in regional centres along important roads (e.g. Gilgit, Leh). Since these locations are easier accessible than most of the villages in this region, they receive imports of meat and other animal products from the lowlands, which is sold in local butcheries (KREUTZMANN 1989, 147). Thus, the availability of animal products in regional centres can be independent of the local system of animal husbandry.

13 The mean precipitation in the valleys' bottom does not exceed 150–200 mm (JACOBSEN 1994, 16–17).

ploughing (oxen) and threshing (donkeys, cows and oxen), which have to be done in spring and autumn, respectively. Thus, once the fields are sown, most of the animals are dispensable in the permanent villages and are taken to the high pastures, where they remain until the harvest season. The spatial division of crop production and animal husbandry serves to prevent animals from entering the fields and eating up the grain. On the other hand, grazing the livestock on the alpine pastures in summer is the only way to make use of ecological niches at higher altitudes, which are not suitable for cultivating cereals and any other kinds of crops. Hence, crop cultivation and animal husbandry are highly interdependent.[14]

The cultivation of the land is most labour-intensive during spring and autumn, i.e. during the preparation of the fields and the harvest, respectively. By contrast, animal husbandry takes up most of the time in summer, when high quantities of milk have to be processed, and when the animals which remain in the homestead have to be kept away from the fields. Thus, the labour peaks are distributed evenly throughout the year, which is usually an advantage for the household labourers. However, due to the prevailing division of work according to age and sex, this advantage is not fully felt in Yasin (HERBERS 1998, 126–142).

A major characteristic of combined mountain agriculture is its high degree of diversification, based on a large variety of crops and species and a wide spatial distribution of fields and grazing grounds over various locations and altitudes. This agricultural diversity results from the necessity to use as many ecological niches as possible. It also has the additional advantage of reducing the risk of being seriously hit by natural hazards.[15] Since it is unlikely that such an event affects all segments of agricultural production simultaneously, the unimpaired areas can act as buffers which secure livelihood even in the case of partial losses in production. Animal husbandry is thus part of an elaborate insurance system.

A peculiarity of High Asia is the small number of roads between the valleys, within the valleys and especially to the summer pastures. Thus, a reliable system of public transport exists only in few areas (e.g. Hunza). As far as Yasin is concerned, there are a few tractors and taxi-jeeps, which operate almost exclusively between the valley and the town of Gilgit. In order to carry heavy loads from one place to another within Yasin, every household keeps one or two donkeys. This kind of transport is of particular importance for the communication between the villages and the high pastures, as the animals are used, for instance, to move equipment (vessels, bedding etc.) between the households in the permanent and the summer villages, or to transport the huge winter supply of firewood from the high pastures to the homesteads.[16]

---

14  Cf. the introductory chapter in this volume.
15  In the autumn of 1992, heavy rain continuing for more than 72 h throughout all of Northern Pakistan caused the death of many people and animals and led to the destruction of buildings, roads and a great proportion of the harvest (BOHLE & PILARDEAUX, 1993).
16  In contrast to Yasin, not donkeys but yaks are used as a means of transportation in other parts of High Asia (BISHOP 1990, 246–247).

## Economic dimensions

The need to minimize the above mentioned risks points to the economic dimensions of animal husbandry. As in other regions of developing countries, the majority of the people in Yasin have only few savings. Thus, livestock is one of the most important capital investments for the high mountain farmers. An ox or a cow which gives milk, for instance, is worth about Rs 4,000 and Rs 3,000, respectively. This capital represents a reserve which farmers can draw on when they are in need of cash. The medical treatment of a household member, the preparation of a wedding, the repayment of debts, a long journey and similar events require that larger sums of money be available. According to the amount of money which a household needs, the number and kind of animals is selected for sale. However, a minimum number of animals is always kept to ensure their reproduction, which is the long-term basis of the capital investment.

Keeping animals as capital investment is not the only economic purpose of animal husbandry. The provision of various raw materials is another important reason why farmers breed livestock. It is men who spin goat and yak hair and weave carpets from it, whereas women spin sheep wool to make pullovers and socks or to get woollen cloth prepared by a local weaver. Moreover, hides from yaks, cattle, goats and sheep are used to make leather sacks for grain and skin containers, which are used to prepare butter (*tárink*). By contrast, the traditional home-made leather shoes have been replaced completely by imports, predominantly of plastic shoes and slippers, from the lowlands and China. A surplus of wool, hair and hides is often marketed. Wool is sold to traders in Gilgit, hides to Pathans from Peshawar who travel through Northern Pakistan to buy these items. However, the proceeds from the sale are usually low, because the farmers sell only small quantities.[17] Other raw materials gained from animal husbandry, such as bones, intestines, chicken feathers and rabbit pelts, are neither processed in the household nor sold on the market, because there is no demand for these raw materials.

Firewood is extremely scarce in Yasin, like in many northern parts of the Hindukush, Karakorum and Himalaya, where large forests could not develop mainly due to the arid environmental conditions. Since the few woods around Yasin have been almost completely cut down, and the market prices for firewood are steadily increasing,[18] a growing number of households which cannot afford kerosene stoves use dung as alternative resource for cooking and heating. However, since manure is also needed to fertilize the fields, women gather dung from roads, along channels and on pastures to supplement the firewood; manure from the livestock sheds, on the other hand, is reserved for the fields. Formerly, animals and people lived in the same building in order to keep the houses warm.

---

17  The main local cash crops include dried apricots, apricot seeds, walnuts, potatoes and *qurút*. Even if these are taken into account, the profits remain rather modest.
18  In 1993, the price for one *man* firewood (approximately 40 kg) amounted to Rs 60. Two years later, in 1995, it had risen to Rs 75.

Although this is an efficient way of heating, this practice has been abandoned for hygienic reasons. Owing to information campaigns of various aid organisations, this method of heating is today no longer in use in Yasin.[19]

## Social and cultural dimensions

The basic social unit in Yasin is the poly-nuclear extended family comprising two or three generations. While the daughters in a family leave the household when they get married, the sons with their wives and children stay with their parents. Thus, a family may comprise up to 30 members living together in one house.[20] Despite its large size, the household depends in many respects on social interaction and co-operation with other households. Therefore, it maintains close relations with paternal relatives (*hápa*), married daughters and sisters of the family (*seléndaru*), foster-families (*úšam*) and neighbours (*girám*). There are also connections with the father's kinship group (*qóom*) and maternal relatives, but these are considered as less important.

This social network is founded on the reciprocal exchange of invitations, labourers and gifts and follows fixed rules. Thus, specific relationships require the exchange of certain kinds of goods. Invitations to festivities and food gifts are the basis of almost every relationship. By contrast, the exchange of labourers and animals as well as gifts other than dishes (money, livestock, bedding etc.) is reserved for particular social units and special occasions. In the following I will give some examples to illustrate that the goods which are exchanged are not always equal in kind, size and value (Fig. 4).

In order to manage the preparation of the fields in spring, most of the households co-operate with their neighbours. To help a household spread the manure on their fields, for instance, each of the co-operating neighbours provides one male labourer. When the time has come for the others to prepare their fields, they are assisted in the same way. Moreover, most households keep only one ox, but two animals are required to plough the fields. Therefore, two neighbours usually co-operate by lending each other their animals. While this is an example where the exchanged goods are equal in kind, the following cases relate to partly unequal exchanges. At the end of the fasting month, during the festival of *ramazán*[21], *paratá* (round flat loaves of bread fried in oil), are circulated in the community (see Fig. 4).[22] Differences in the value of these gifts result from the

---

19  In order to protect lambs and calves against low outside temperatures, they are still kept in the house. The same holds true for chicken, for which adequate sheds are usually missing.
20  Households are divided when the extended family becomes too large to live in one house, when the head of the family, i.e. the father, passes away, or when a male household member finds employment outside the valley and migrates with his wife and children to a new location.
21  Another common name for this festival is ʿid ul-fitr.
22  Other occasions to send dishes to relatives and neighbours include the festival at the beginning of the field work (*bo*), the return of the household members from the summer pastures (*térum dosókićum*), and weddings (see Fig. 2).

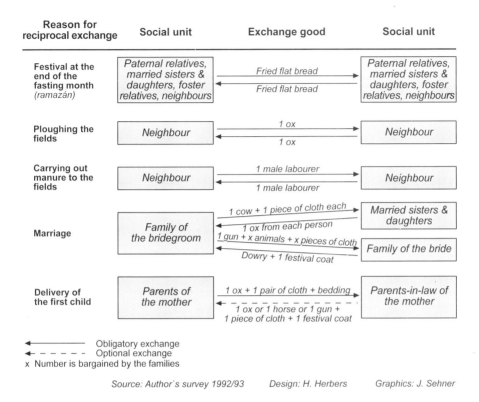

Fig. 4: Examples of reciprocal exchange in Yasin

number of loaves given and the kind of oil which was used for their preparation. A wealthy household may present their relatives with *paratá* prepared with walnut oil, while the recipients may return bread fried in oil which is of cheaper quality, because they cannot afford high-quality oil. The same holds true for the arrangement of large festivities, such as weddings and funerals, for which invitations are exchanged. Meat is the most important dish which has to be served to the guests on these occasions. Which and how many animals are slaughtered depends significantly on the wealth of the family, i.e., on the number of livestock they keep, or alternatively, on their monetary budget to purchase animals.

Finally, there is a third category of reciprocal relationships, in which the exchanged goods have nothing in common at all. This is true, for instance, for the mutual obligations between two families whose children are married to each other. On this occasion, the family of the bridegroom has to present the bride's family with several animals, some pieces of cloth and a gun, whereas the bridegroom's sisters each get one cow and a piece of cloth. In return, the bride's family presents the groom's parents with the daughter's dowry and a *jumú*, a local ceremonial coat, and each of the sisters give one ox to the bridegroom's family.[23] Comparable exchanges (Fig. 4) take place when a woman has delivered her first child, irrespective of whether it is a son or a daughter. However, in this case her in-laws are not obliged to reciprocate the gifts to her parents.

The main function of such reciprocal exchanges is to maintain good social relations. These are not only important in order to manage agricultural tasks and to guarantee the well-being of married daughters and sisters in the families of their husbands. Good social relations are particularly important in times of crisis. The death of the head of a family, high debts, the destruction of a family's house during an earthquake or other natural disasters may cause critical situations,[24] in which the affected household needs the support of relatives and neighbours. Since many mutual exchanges rely on animals or animal products, the availability of livestock is crucial. Households which have only few animals may therefore be at a disadvantage, especially because it is unusual to give money or other goods in exchange for an animal; furthermore, even if this was common practice, most of the households would not be in a position to give money.

Another social dimension of animal husbandry relates to the local conception of hospitality. As mentioned earlier, the host is expected to entertain a high guest to milk tea, meat, eggs, butter or other dairy products. The ability to supply sufficient animal products on such occasions depends on the number of animals the host keeps.

In Yasin, animals are also used for sacrificial purposes, both with respect to non-Islamic and Islamic ceremonies. By way of example, a house in which a woman has delivered a child has to be 'purified' in order to soothe the fairies, in which the local population believes. This is done by burning juniper branches and sacrificing a chicken, which is then eaten by the midwife and the female members of the household. Furthermore, the celebration of *qorbaní*, one of the highest festivals for Muslims (see Fig. 2), requires the slaughtering of a goat, as it celebrates Ibrahim's readiness to sacrifice his only son to god. Since the people in Yasin keep comparatively few animals, only those goats are slaughtered on *qorbaní* which are born on a particular date. Thus, usually only one household in a hamlet sacrifices a *qorbaní*-goat and then shares its meat with relatives and neighbours (see Photo 2).

While animals are generally kept for a variety of reasons, horses are kept in Yasin for the single purpose of playing polo, a very famous sport which has a long tradition in this area. Once a year a polo tournament is held on the Shandur Pass (3,600 m) in Northern Pakistan, which attracts many visitors from all over the country. Although the polo players from Yasin are not professional enough and their horses in too poor a condition to take part in this important event, some

---

23 The exact number of animals and clothes to be exchanged as well as the dowry's value is bargained by representatives of the bridegroom's and the bride's family.
24 Earthquakes are common in High Asia. In Yasin, minor seismic shocks occur several times every year. The last serious earthquake in the valley, which caused heavy damages and killed several people, was recorded in 1943 (eight points on the Richter scale, cf. HUGHES 1984, 272–274).

Photo 2: A *qorbaní*-goat is prepared for slaughtering (Sandi, June 1992)

wealthy people and those who are passionately fond of polo keep a horse exclusively for local matches and against teams from adjacent valleys. Since cultural events are rare in this mountain area, polo is highly appreciated by all people of the local community.

## 5. ANIMAL VERSUS PLANT PRODUCTS FOR A HEALTHY DIET

In many cultures, animal products, and especially meat, have a special status. The high value of these food items cannot only be related to their particular taste but also to their limited availability and high market prices. This holds true for the present situation in many developing countries as well as for western countries before World War II. In addition, animal products are generally considered as healthier than plant products – a belief which is not corroborated by scientific findings, as the following paragraphs will show.

The analysis of the nutrient composition of animal products shows that they are particularly rich in protein, iron, calcium and riboflavin, and, in the case of butter, also in fat (Tab. 2). Deficiencies in these nutrients cause various physiological disorders and diseases. However, it is not only animal products but also several plants which contain high amounts of the above mentioned nutrients. Except for vegetable oil, which can be used as a substitute for butter, all plant products listed in Tab. 2 are quite common in the diet of the farmers in Yasin. Moreover, the energy provided by the fat of butter can also be supplied through carbohydrates. Consequently, if bread, noodles and other foodstuff rich in carbohydrates are consumed in sufficient quantities, high intakes of fat are not necessary. By contrast, iron from animal products is nutritionally more valuable than plant iron, because the former has a higher absorption rate, the so-called bioavailability. However, nutritionists do not agree as to whether the lower bioavailability of plant iron is overcompensated by its higher total proportions in cereals and vegetables (HALLBERG, SANDSTRÖM, AGGETT 1993, 179).[25] Like iron, the protein in animal products is of higher nutritional value than that in plants, since it contains high proportions of essential amino acids.[26] Thus, the protein efficiency ratio, i.e. the capability to be converted into protein tissues of the human body, is higher in animal products.[27] Yet, as long as the energy requirements of a person are not met – and this is a basic problem in developing countries including High Asia (HERBERS 1998, 214–250) – protein is exclusively used as a source of energy, irrespective of its origin. It is thus more important to improve the energy supply than the quantity and quality of the protein.

Another aspect regarding the nutritional value of animal products has to be mentioned. While eggs, meat, milk, cottage cheese etc. are rich in some nutrients, they are poor in others (e.g. Vitamin A and C) which can be found in high concentrations in plant products. Since the nutrients contained in cereals, vegetables, fruits etc. are as essential as those in animal products, it is not justified to consider the latter as healthier and better than plant products. Thus, a balanced diet should comprise a large diversity of foodstuff, i.e. ideally, a combination of both animal *and* plant products. Hence, the problem in Yasin is not the vegetarian diet as such, but the monotone composition of the food intake. A diet, which consists mainly of wheat bread with tea, and which is deficient in energy, can hardly provide all the essential nutrients. Thus, at present the nutritional value of the few animal products which the people of Yasin consume consists mainly in adding flavour to the meals and occasionally breaking up the monotonous diet. Since the nutritional and health status of the local farmers is insufficient (HERBERS

---

25  Moreover, the absorption of iron is a very complex process, influenced in various ways by other food components. By way of example, phytates in tea and calcium in milk inhibit the absorption of iron, while ascorbic acid or meat enhance it. In addition, iron deficiency results not only from an unbalanced food intake but also from diarrhoea, other intestinal diseases and worm infections, which cause reduced absorption rates and sometimes the loss of blood.

26  Essential nutrients are those which the organism can not synthesize itself (e.g. vitamins, minerals, eight amino acids, two fatty acids). These elements must be provided with the food.

27  However, it is possible to improve the amino acid composition of plant products and therefore to enhance the protein efficiency ratio by combining specific food items. Potatoes consumed together with eggs, wheat with milk, maize with beans and other cereals with pulses are very efficient combinations. Hence, a diet poor in meat can provide sufficient high quality protein if the diet has the right composition. Unfortunately, such dietetic knowledge is usually not very widespread.

1998, 233–250), the level of energy and nutrients in their diet needs to be increased through the provision of larger quantities and a greater variety of plant and animal products. Although cereals, vegetables and fruits could substitute for meat, eggs, milk and dairy products to some extent, the latter are definitely high quality products. Except for meat, animal products are indispensable for a balanced diet which meets the general needs of the mountain dwellers and, more particularly, the additional needs of children, pregnant and breastfeeding women, elderly or sick inhabitants. The introductory statement that animal husbandry is not reflected in the local diet, has thus further implications. Since an adequate diet is a prerequisite for growth, mental development, health and working capacity, the imbalance between the economic system and the nutritional patterns proves to be a significant disadvantage for the local community.

Tab. 2: Functions of major nutrients in animal products, effects of deficiencies and alternative intake sources from plant products

| Main physiological functions | Main effects of deficiencies | Animal sources | Plant sources |
|---|---|---|---|
| *Protein* Constituent of all cells and tissues, essential part in enzymes, hormones and antibodies | Reduced wound healing and immune response, anaemia, retardation in growth and mental development, reduced physical working capacity | Eggs, meat | Pulses, potatoes |
| *Fat* Energy supply, energy reserves, constituent of cell membranes | Metabolic disorders only in case of deficiencies in essential fatty acids | Butter | Vegetable oil |
| *Iron* Constituent of haemoglobin and myoglobin | Fatigue, weakness, anaemia, reduced immune response | Eggs, meat, blood | Wheat, fruits, vegetables |
| *Calcium* Constituent of bones and teeth, permeability of cell membranes | Rickets, osteomalacia, osteoporosis | Milk, dairy products | Green vegetables |
| *Riboflavin* Oxidation-reduction reactions | Rash, unspecific symptoms | Eggs, meat, milk, dairy products | Wholemeal products |

Source: ELMADFA & LEITZMANN 1990; GARROW & JAMES, 1993

## 6. FARMERS AS VEGETARIANS: BOWING TO NECESSITY

An increasing number of people in western countries are becoming vegetarians. Some of them are critical of animal products due to bovine spongiform encephalopathy (BSE) and swine fewer, others avoid the consumption of these products due to the practice of intensive livestock farming, which they consider as cruel, or to ethical principles prohibiting the killing of any living being. While the majority of the western vegetarians do not eat meat, they do consume eggs, milk, dairy products, honey and even fish. Whatever their habits, people in western countries are usually free to choose whether they want to live on a vegetarian diet, or which animal products they do not want to eat. By contrast, the mountain farmers in Yasin are mainly vegetarians – or even vegans – against their will. If they could, most of them would eat more frequently, and bigger portions of meat and other animal products. Independent of the fact that animal husbandry has many other important functions than just providing animal products for consumption, it is evident that local farmers would highly appreciate an improved supply of the latter. Yet, the seasonal shortage of fodder resources in this high mountain area prevent them from keeping more livestock. It is to be expected, however, that the desire for more meat, eggs, milk etc. will further increase owing to social changes, such as the growing number of households with incomes from off-farm employment, the rising number of people with higher education or the fact that more people will be exposed temporarily to the amenities of life in the city. As long as this demand can not be met by imports, vegetarianism, along with animal husbandry, will continue to be a significant characteristic of Yasin and many high mountain regions in Asia.

## 7. REFERENCES

ATTENBOROUGH, R., ATTENBOROUGH, M., LEEDS, A.R. (1994): Nutrition in Stongde. In: CROOK, J., OSMASTON, H. (Eds.): Himalayan Buddhist Villages. Environment, Resources, Society and Religious Life in Zangskar, Ladakh. Bristol, 383–404.

ALLAN, N.J.R. (1990): Household Food Supply in Hunza Valley, Pakistan. In: Geographical Review 80, 399–415.

BISHOP, B.C. (1990): Karnali under Stress. Livelihood Strategies and Seasonal Rhythms in a Changing Nepal Himalaya. Geography Research Paper, 228–229. Chicago.

BOHLE, H.-G., PILARDEAUX, B. (1993): Jahrhundertflut in Pakistan, September 1992: Chronologie einer Katastrophe. In: Geographische Rundschau 45, 124–126.

CASIMIR, M.J. (1991): Flocks and Food. A Biocultural Approach to the Study of Pastoral Foodways. Kölner Ethnologische Mitteilungen, 10. Köln.

CASIMIR, M.J. (1998): Pastoralnomadische Strategien und ausgewogene Ernährung: Zwei Beispiele aus Südasien. In: BOHLE, H.-G. et al. (Eds.): Ernährungssicherung in Südasien. Siebte Heidelberger Südasiengespräche. Beiträge zur Südasienforschung, 178, 45–57.

ELMADFA, I., LEITZMANN, C. (1990): Ernährung des Menschen. Stuttgart.

GARROW, J.S., JAMES, W.P.T. (Eds.) (1993): Human Nutrition and Dietics. Edinburgh.

HASERODT, K. (1989): Chitral (pakistanischer Hindukusch). Strukturen, Wandel und Probleme eines Lebensraumes im Hochgebirge zwischen Gletschern und Wüste. In: HASERODT, K. (Ed.): Hochgebirgsräume Nordpakistans im Hindukusch, Karakorum und Westhimalaya.

(Beiträge und Materialen zur Regionalen Geographie 2). Berlin, 44–180.

HALLBERG, L., SANDSTRÖM, B., AGGETT, P.J. (1993): Iron, Zinc and Other Trace Elements. In: GARROW, J.S.; JAMES, W.P.T. (Eds.): Human Nutrition and Dietics. Edinburgh, 208–238.

HERBERS, H. (1998): Arbeit und Ernährung in Yasin. Aspekte des Produktions-Reproduktions-Zusammenhangs in einem Hochgebirgstal Nordpakistans. Erdkundliches Wissen, 123. Stuttgart.

HUGHES, L.E. (1984): Yasin Valley: The Analysis of Geomorphology and Building Types. In: MILLER, K.J. (Ed.): The International Karakoram Project. (Vol.2). Cambridge, 253–288.

JACOBSEN, J.-P. (1994): Climatic Records from Northern Pakistan (Yasin Valley). In: CAK News Letter 2, 13–18.

KREUTZMANN, H. (1989): Hunza: Ländliche Entwicklung im Karakorum. Abhandlungen – Anthropogeographie, 44. Berlin.

LEWIS, B., PELLAT, C., SCHACHT, J. (Eds.) (1965): Encyclopaedia of Islam, Vol. 2. Leiden.

MÜLLER-STELLRECHT, I. (1979–1980): Materialien zur Ethnographie Dardistans (Pakistan). (Part 1: Hunza; Part 2/3: Gilgit. Chitral und Yasin). Bergvölker im Hindukusch und Karakorum, 3. Graz.

SAUNDERS, F. (1983): Karakoram Villages. An Agrarian Study of 22 Villages in the Hunza, Ishkoman and Yasin Valleys of Gilgit District. FAO Integrated Rural Development Project PAK 80/009. Gilgit.

## ADDRESSES OF CONTRIBUTORS

Prof. Dr. Eckart Ehlers, Institut für Wirtschaftsgeographie, Universität Bonn, Meckenheimer Allee 166, D-53115 Bonn

Dr. Georg Stöber, Georg-Eckert-Institut für internationale Schulbuchforschung, Cellerstr. 3, D-38114 Braunschweig

Dr. Hiltrud Herbers, Institut für Geographie, Universität Erlangen-Nürnberg, Kochstr. 4/4, D-91054 Erlangen

Reinhard Fischer, Christstr. 5, D-14059 Berlin

Dr. Marcus Nüsser, Geographische Institute, Universität Bonn, Meckenheimer Allee 166, D-53115 Bonn

Prof. Dr. Hermann Kreutzmann, Institut für Geographie, Universität Erlangen-Nürnberg, Kochstr. 4/4, D-91054 Erlangen

Matthias Schmidt, Institut für Wirtschaftsgeographie, Universität Bonn, Mekkenheimer Allee 166, D-53115 Bonn

Dr. Jürgen Clemens, Institut für Wirtschaftsgeographie, Universität Bonn, Mekkenheimer Allee 166, D-53115 Bonn

51. **Helmut J. Jusatz, Hrsg.: Geomedizin in Forschung und Lehre.** Beiträge zur Geoökologie des Menschen. Vorträge des 3. Geomed. Symposiums auf Schloß Reisensburg vom 16. - 20. Okt. 1977. Hrsg. im Auftrag der Heidelberger Akademie der Wissenschaften. 1979. XV, 122 S. m. 15 Abb. u. 14 Tab., 1 Faltkte., Summaries, kt. 2801-3
52. **Werner Kreuer: Ankole.** Bevölkerung - Siedlung - Wirtschaft eines Entwicklungsraumes in Uganda. 1979. XI, 106 S. m. 11 Abb., 1 Luftbild auf Falttaf., 8 Ktn., 18 Tab., kt. 3063-8
53. **Martin Born: Siedlungsgenese und Kulturlandschaftsentwicklung in Mitteleuropa.** Gesammelte Beiträge. Hrsg. im Auftrag des Zentralausschusses für Deutsche Landeskunde von Klaus Fehn. 1980. XL, 528 S. m. 17 Abb., 39 Ktn. kt. 3306-8
54. **Ulrich Schweinfurth / Ernst Schmidt-Kraepelin / Hans Jürgen von Lengerke / Heidrun Schweinfurth-Marby / Thomas Gläser / Heinz Bechert: Forschungen auf Ceylon II.** 1981. VI, 216 S. m. 72 Abb., kt. (Bd. I s. Nr. 27) 3372-6
55. **Felix Monheim: Die Entwicklung der peruanischen Agrarreform 1969-1979 und ihre Durchführung im Departement Puno.** 1981. V, 37 S. m. 15 Tab., kt. 3629-6
56. **- / Gerrit Köster: Die wirtschaftliche Erschließung des Departement Santa Cruz (Bolivien) seit der Mitte des 20. Jahrhunderts.** 1982. VIII, 152 S. m. 2 Abb. u. 12 Ktn., kt. 3635-0
57. **Hans Georg Bohle: Bewässerung und Gesellschaft im Cauvery-Delta (Südindien).** Eine geographische Untersuchung über historische Grundlagen und jüngere Ausprägung struktureller Unterentwicklung. 1981. XVI, 266 S. m. 33 Abb., 49 Tab., 8 Kartenbeilagen, kt. 3550-8
58. **Emil Meynen / Ernst Plewe, Hrsg.: Forschungsbeiträge zur Landeskunde Süd- und Südostasiens.** Festschrift für Harald Uhlig zu seinem 60. Geburtstag, Band 1. 1982. XVI, 253 S. m. 45 Abb. u. 11 Ktn., kt. 3743-8
59. **- / -, Hrsg.: Beiträge zur Hochgebirgsforschung und zur Allgemeinen Geographie.** Festschrift für Harald Uhlig zu seinem 60. Geburtstag, Band 2. 1982. VI, 313 S. m. 51 Abb. u. 6 Ktn., 1farb. Faltkte., kt. 3744-6
Beide Bände zus. kt. 3779-9
60. **Gottfried Pfeifer: Kulturgeographie in Methode und Lehre.** Das Verhältnis zu Raum und Zeit. Gesammelte Beiträge. 1982. XI, 471 S. m. 3 Taf., 18 Fig., 16 Ktn., 15 Tab. u. 7 Diagr., kt. 3668-7
61. **Walter Sperling: Formen, Typen und Genese des Platzdorfes in den böhmischen Ländern.** Beiträge zur Siedlungsgeographie Ostmitteleuropas. 1982. X, 187 S. m. 30 Abb., kt. 3654-7
62. **Angelika Sievers: Der Tourismus in Sri Lanka (Ceylon).** Ein sozialgeographischer Beitrag zum Tourismusphänomen in tropischen Entwicklungsländern, insbesondere in Südasien. 1983. X, 138 S. m. 25 Abb. u. 19 Tab., kt. 3889-2
63. **Anneliese Krenzlin: Beiträge zur Kulturlandschaftsgenese in Mitteleuropa.** Gesammelte Aufsätze aus vier Jahrzehnten, hrsg. von H.-J. Nitz u. H. Quirin. 1983. XXXVIII, 366 S. m. 55 Abb., kt. 4035-8
64. **Gerhard Engelmann: Die Hochschulgeographie in Preußen 1810-1914.** 1983. XII, 184 S., 4 Taf., kt. 3984-8
65. **Bruno Fautz: Agrarlandschaften in Queensland.** 1984. 195 S. m. 33 Ktn., kt. 3890-6
66. **Elmar Sabelberg: Regionale Stadttypen in Italien.** Genese und heutige Struktur der toskanischen und sizilianischen Städte an den Beispielen Florenz, Siena, Catania und Agrigent. 1984. XI, 211 S. m. 26 Tab., 4 Abb., 57 Ktn. u. 5 Faltktn., 10 Bilder auf 5 Taf., kt. 4052-8
67. **Wolfhard Symader: Raumzeitliches Verhalten gelöster und suspendierter Schwermetalle.** Eine Untersuchung zum Stofftransport in Gewässern der Nordeifel und niederrheinischen Bucht. 1984. VIII, 174 S. m. 67 Abb., kt. 3909-0
68. **Werner Kreisel: Die ethnischen Gruppen der Hawaii-Inseln.** Ihre Entwicklung und Bedeutung für Wirtschaftsstruktur und Kulturlandschaft. 1984. X, 462 S. m. 177 Abb. u. 81 Tab., 8 Taf. m. 24 Fotos, kt. 3412-9
69. **Eckart Ehlers: Die agraren Siedlungsgrenzen der Erde.** Gedanken zu ihrer Genese und Typologie am Beispiel des kanadischen Waldlandes. 1984. 82 S. m. 15 Abb., 2 Faltktn., kt. 4211-3
70. **Helmut J. Jusatz / Hella Wellmer, Hrsg.: Theorie und Praxis der medizinischen Geographie und Geomedizin.** Vorträge der Arbeitskreissitzung Medizinische Geographie und Geomedizin auf dem 44. Deutschen Geographentag in Münster 1983. Hrsg. im Auftrage des Arbeitskreises. 1984. 85 S. m. 20 Abb., 4 Fotos u. 2 Kartenbeilagen, kt. 4092-7
71. **Leo Waibel †: Als Forscher und Planer in Brasilien:** Vier Beiträge aus der Forschungstätigkeit 1947-1950 in Übersetzung. Hrsg. von Gottfried Pfeiffer u. Gerd Kohlhepp. 1984. 124 S. m. 5 Abb., 1 Taf., kt. 4137-7
72. **Heinz Ellenberg: Bäuerliche Bauweisen in geoökologischer und genetischer Sicht.** 1984. V, 69 S. m. 18 Abb., kt. 4208-3
73. **Herbert Louis: Landeskunde der Türkei.** Vornehmlich aufgrund eigener Reisen. 1985. XIV, 268 S. m. 4 Farbktn. u. 1 Übersichtskärtchen des Verf., kt. 4312-8
74. **Ernst Plewe / Ute Wardenga: Der junge Alfred Hettner.** Studien zur Entwicklung der wissenschaftlichen Persönlichkeit als Geograph, Länderkundler und Forschungsreisender. 1985. 80 S. m. 2 Ktn. u. 1 Abb., kt. 4421-3
75. **Ulrich Ante: Zur Grundlegung des Gegenstandsbereiches der Politischen Geographie.** Über das "Politische" in der Geographie. 1985. 184 S., kt. 4361-6
76. **Günter Heinritz / Elisabeth Lichtenberger, eds.: The Take-off of Suburbia and the Crisis of the Central City.** Proceedings of the International Symposium in Munich and Vienna 1984. 1986. X, 300 S. m. 95 Abb., 49 Tab., kt. 4402-7
77. **Klaus Frantz: Die Großstadt Angloamerikas im Wandel des 18. und 19. Jahrhunderts.** Versuch einer sozialgeographischen Strukturanalyse anhand ausgewählter Beispiele der Nordostküste. 1987. 200 S. m. 32 Ktn. u. 12 Abb. kt. 4433-7
78. **Claudia Erdmann: Aachen im Jahre 1812.** Wirtschafts- und sozialräumliche Differenzierung einer frühindustriellen Stadt. 1986. VIII, 257 S. m. 6 Abb., 44 Tab., 19 Fig., 80 Ktn., kt. 4634-8
79. **Josef Schmithüsen †: Die natürliche Lebewelt Mitteleuropas.** Hrsg. von Emil Meynen. 1986. 71 S. m. 1 Taf., kt. 4638-8
80. **Ulrich Helmert: Der Jahresgang der Humidität in Hessen und den angrenzenden Gebieten.** 1986. 108 S. m. 11 Abb. u. 37 Ktn. i. Anh., kt. 4630-5
81. **Peter Schöller: Städtepolitik, Stadtumbau und Stadterhaltung in der DDR.** 1986. 55 S., 4 Taf. m. 8 Fotos, 12 Ktn., kt. 4703-4